MW01366488
book's for you
Jeff

A GREAT UNDERTAKING

A GREAT UNDERTAKING

Mechanization and Social Change in a Late Imperial Chinese Coalmining Community

Jeff Hornibrook

Published by State University of New York Press, Albany

Printed in the United States of America

For information, contact State University of New York Press, Albany, NY
www.sunypress.edu

Production, Diane Ganeles
Marketing, Anne M. Valentine

Library of Congress Cataloging-in-Publication Data

Hornibrook, Jeff.
A great undertaking : mechanization and social change in a late Imperial Chinese coalmining community / Jeff Hornibrook.
pages cm
Includes bibliographical references and index.
ISBN 978-1-4384-5687-4 (hardcover : alk. paper)
ISBN 978-1-4384-5689-8 (e-book : alk. paper)
1. Coal mines and mining—China—Pingxiang (Jiangxi Sheng)—History—19th century. 2. Industrialization—China—Pingxiang (Jiangxi Sheng)—History—19th century. 3. Pingxiang (Jiangxi Sheng, China)—Economic conditions—19th century. 4. Pingxiang (Jiangxi Sheng, China)—Social conditions—19th century. 5. Pingxiang (Jiangxi Sheng, China)—History—19th century. I. Title.

HD9556.C53P564 2015
338.2'7240951222—dc23 2014030656

10 9 8 7 6 5 4 3 2 1

Contents

Acknowledgments

While sitting in my office tracking down Chinese characters for the glossary and figuring out rent calculations in piculs of rice, it is easy to think that this book was done by my own hand with little outside assistance. In reality, this project has taken a very long time and relied on the kindness of many friends, colleagues, family members—and some strangers—along the way.

First, I would like to thank my professors at the University of Minnesota who helped me through the initial stages of this project. Because they were often skeptical that a study of modern coalmining was a suitable topic for a dissertation, I was doubly aware that my work had to pass the smell test on a number of occasions. Even more, for more than two years Drs. Romeyn Taylor and Ann Waltner took time out of their days to help me with my translations and think through the implications of the data I collected. Along with Ted Farmer, they also guided me through the initial process of learning to do research using Chinese sources, a skill that does not simply translate from American library systems.

Once I completed my dissertation, I found new data that dramatically altered my understanding of the history of Pingxiang County. I honestly didn't know what to do with the increasingly apparent notion that Confucian scholars were also running firms as big as Guangtaifu. Fortunately, I could turn to an old friend, David Wakefield, whose understanding of Chinese history was unique and insightful. He set me straight and cleared up many misperceptions I had about my own data. Sadly, with his death, I lost a sounding board who listened to me and talked me through problems big and small. More importantly, I lost a true friend.

Translating and analyzing documents is, to say the least, a daunting task. I turned to many Chinese scholars and students to help me with translations and broader understanding of the documents on the social history of coalmining in Pingxiang County. For this endeavor I would like to thank Yao Yusheng, Hsu Pi-ching, Jiang Yonglin, Pan Ming-te, Zhu Lisheng, Hao Yongjuan, Leigh Zhang, Zhang Jiefu, Stephanie Balfoort, Liu Tongfei, Li Minmin, Du Yisi, and Lily Wang. Also, a big thanks to Dr. Xie Liou for making the map of Pingxiang County and the environs.

Most of the research funding for this project came from State University of New York at Plattsburgh, where I have taught for the past sixteen years. Two presidential research scholarships allowed me to travel to Pingxiang County and find documents I could not find elsewhere and to simply take in the sights and feel of Anyuan and Pingxiang County. This work was greatly assisted by the wonderful staff at the Jiangxi Provincial Academy of Social Sciences in Nanchang and the branch office in Pingxiang County. To the many people in those two offices: I cannot thank you enough.

From the time I completed this work as a dissertation until now I have been adding/tweaking/rethinking much of the material as I put together papers and articles. I was inspired and assisted by my colleagues in the History Department at SUNY Plattsburgh, including Vincent Carey, Wendy Gordon, Gary Kroll, Stuart Voss, Doug Skopp, Jim Lindgren, Richard Schaefer, Jim Rice, Ryan Alexander, Mark Richard, and Sylvie Beaudreau. Also, while attending conferences, I met and gained insights from Elizabeth Perry, Madeline Zelin, Andrea McElderry, Joseph Esherick, Chris Isett, Fa-ti Fan, and Kristen Stapleton. However, it was not until I sent my draft to State University of New York Press that this research began to take on the form of a real book manuscript. The press's anonymous readers provided me with sharp criticisms that allowed me to take my data and formulate it into something more organized and complete. I subsequently joined a writing group on campus made up of Monica Ciobanu, Dan Lake, Connie Shemo, and Jessamyn Neuhaus. They read each chapter and helped me bring my pages into book form. I must especially thank Connie Shemo for pushing me to make contacts and put myself out to the scholarly community when my personal inclination was to draw back. And I cannot miss the opportunity to thank Jessamyn Neuhaus for going over my text with a fine-toothed comb and encouraging me to make stylistic changes that made the narrative clearer and much livelier (how can I ever repay her for all this?).

Lastly, I wish to thank all my family members who have waited for a very long time to see me complete this work. In many ways, I received support and respite from each of them. I especially wish to acknowledge my beautiful daughter Olivia and my lovely wife Annette. If it wasn't for them there would be no reason to strive to succeed, or at least success would have no payoff. It is to those two wonderful women in my life that this book is dedicated.

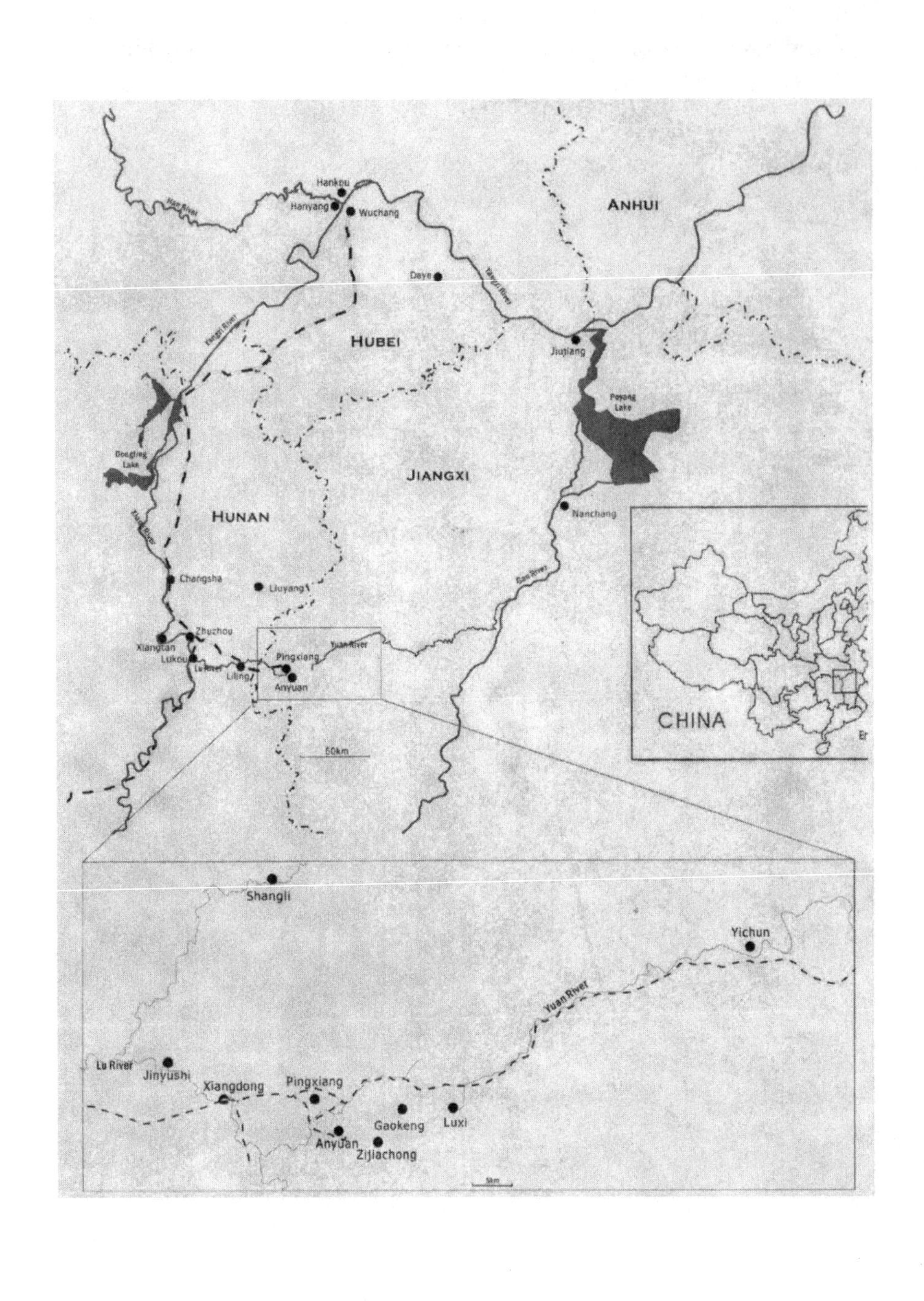

Hankou
Hanyang
Wuchang
Han River
ANHUI
Daye
Yangzi River
HUBEI
Jiujiang
Poyang Lake
Dongting Lake
JIANGXI
Nanchang
HUNAN
Gan River
Changsha
Liuyang
Zhuzhou
Xiangtan
Lukou
Liling
Pingxiang
Yuan River
Anyuan
50km
CHINA
Shangli
Yichun
Yuan River
Lu River
Jinyushi
Xiangdong
Pingxiang
Gaokeng
Luxi
Anyuan
Zijiachong
5km

Introduction

In the spring of 1896, the people of Pingxiang County pursued their lives as they had for generations, indeed for centuries. In this isolated community located in Jiangxi Province just along the provincial border with Hunan to the west, peasants began the backbreaking work of planting their landlords' rice fields and then continued throughout the summer with weeding and maintaining the crop. When the summer crop was harvested, they put in a fall crop of wheat or sweet potatoes, and then, when the winter months made farming no longer possible, the men and boys went off to the mountains to mine for coal. Using essentially the same tools they employed for farming—as these were the only ones they owned—they headed off to the landlords' coalfields in hopes that they could extract pieces of coal for heating their homes, fueling their ovens, and perhaps selling in the market for a few extra coins. They dug holes in the mountains marked by small flags and extracted the minerals until the vein was depleted or the water table flooded the shaft. Then the work teams moved to the next spot, often a few yards away; planted another flag; and began the process again.

That fall, however, a German miner went into the mountains and began examining the quality of the minerals. The man's name was Gustav Leinung, and he was sent by the imperial court to examine the coal deposits to determine if they could be used to fuel an ironworks in the Wuhan Cities to the north. This new factory was to be the focal point of a modernization scheme beyond the wildest dreams or understanding of most of the people of the county. And indeed, twenty years later, Pingxiang County's mountains around the mining town of Anyuan were covered not with small mine shafts and flags but with railroads and factories, European compounds and workers' dormitories, Christian churches and hospitals, gambling houses and opium dens. The changes were not only architectural, however. Local elites and lineage leaders that had overseen the politics and economy of Pingxiang County for centuries were pushed aside and replaced by a new sociopolitical blueprint. Mining was not only done seasonally but was year-round, and local peasants were increasingly working side-by-side with men recruited both regionally and even throughout the greater Chinese empire. Perhaps most importantly, the extracted coal came out of the mines in quantities one hundred times the amount it had in the past. And when it was taken from the mountains, it was placed on a modern train and sent far away from the county's markets and the inhabitants that had relied on it for generations. The coal shipments traveled several hundred miles to factories that were forged to modernize the economy of cities and markets most people in Pingxiang County would never see.

This book is a micro history of the social and political effects of industrialization in Pingxiang County, a Chinese coalmining community. Instead of focusing on entrepreneurs and managers interacting within a bureaucratic hierarchy or looking at workers inside union halls, this study examines the greater community in which the mine was nestled and shows that the mechanization process of Pingxiang County came about through negotiation and struggle that altered the land and society in unforeseen and disruptive ways. Ultimately, as they negotiated the terms of mineral extraction, these people simultaneously struggled over new designations of each other's class and status in the county and the empire. They not only had to redefine their friends and enemies, superiors and subordinates, but they began to see themselves in a new light, altering their class identifications and even, perhaps, consciousness. Thus, when the dust finally settled, not only was the coalmining scheme reorganized, but nearly everyone who touched this project, from top to bottom, had been reclassified as well. Specifically, the county magistrate became a purchase agent of paddy fields in the interests of the imperial state and to the detriment of the needs of the county's landholders and gentry families. The local gentry that owned many of the county's mines were marginalized by a newly emerging managerial class whose personal connections more closely linked them to Chinese compradors and even foreign merchants. Local students became more political and nationalistic as foreigners arrived to reorganize production and transportation systems that had been in place in the county for centuries. Commoners, including miners and boat haulers, saw their family members' occupations subsumed within the larger scheme, while new technologies dramatically changed their daily activities and social status from petty service industry workers to laborers. This study of a coalmine in a county in south central China is not simply an examination of an isolated mining town. Rather, it is a case study of the dramatic global forces that reordered societies and governments, families and communities, all over the globe as they were experienced in a remote part of the world.

Understanding the impact of modernization on a rural Chinese community is significant in part because most historians who examine modernization do not focus on the community but rather on one of two other broad methodologies. First, some business historians interested in international business or foreign trade place the foundations of various mining enterprises within the context of entrepreneurial or governmental development. This research often utilizes memos between managers, corporate ledgers, or bank contracts to see the inner workings of the corporation. Scholars of this type of study, especially those interested in the "Why the West succeeded and China failed" question, focus on the ability or failure of Western capital to create modern firms, or they examine the political and economic struggles that went into modernizing non-Western corporations.[1] In Chinese history, Albert Feuerwerker, Wellington Chan, and Elisabeth Köll examine industrialization projects of various types and essentially argue that the conflict between Confucian ideology and Chinese values and Western technology led to a series of failed attempts at modernization.[2] While these studies have provided reams of data on the industrial schemes and placed some of that data into the context of modern capitalist industrialization, they fail to see the outcomes of their actions on the lives of workers, consumers, or the general population. Managers are viewed making decisions that either work or fail but not as considering the impact those outcomes had on the people who were ordered to implement them. Authors seemingly place the corporations within an isotropic plain devoid of a local society, economy, or history.

More recently, labor histories interested in nineteenth-century industrialization discuss the creation of a nascent proletariat in factories or cities. These studies, often based on the Marxist theories of E. P. Thompson and others, focus on the daily material lives of workers and on the role of class and the rise of unions or other labor activities.[3] In all, they point to a significant break with the past as mechanization brought capital and labor into the process of production. For mining specifically, technical advancement changed the daily lives of people as seasonal laborers became year-round workers and migrant workers became organized laborers. In the case of Pingxiang County coalmining, monographs by Lynda Shaffer and Elizabeth Perry examine the history of the labor movement in the mining town of Anyuan during the Republican era. They both point out that the coalminers were initially organized under the clientage system of contract labor bosses that was gradually replaced in the early 1920s by a modern labor union organized by the Chinese Communist Party.[4] While these studies bring the reader into the mining town, they ignore the gentry and peasants who also lived in the surrounding community as it was incorporated into a modern world few had previously imagined. In so doing, they failed to see the relationships that were altered and fostered as mechanization transformed the economy of the county as a whole almost as much as it affected the mining town.

If Western scholarship on modernization is incomplete and at least partially based on political interests of labor and capital, Chinese scholarship on the history of Pingxiang County and even mining in general is even more fragmented and is almost always oriented toward proving the current ideology of the Communist Party leadership. No monographs have been written by Chinese scholars on the mines in Pingxiang County. However, the available literature indicates that since the Communist government took power in the late 1940s, the Chinese have moved in the opposite direction from Western scholarship beginning with revolutionary tracts and moving eventually to studies of markets and technological development. Early sources focus primarily on the Communist Party's activities and the swelling support of the workers in the mines. Most of this material can be classified as little more than propaganda published to exalt the leadership of Mao Zedong and, to a lesser extent, his subordinate and future heir apparent, Liu Shaoqi.[5] However, with the deaths of both of these men and the rise of the pro-West, pro-modernization themes of the current era, new studies have been published that examine China's past economic successes and point to possible avenues of development for the future. Scholars from the Chinese Academy of Social Sciences examine the history of China as it moved from a "natural economy" to a more modern system that signified the "sprouts of capitalism." They attempted to show that some Chinese merchants improved the forces of production through factories that were larger than previous endeavors in China and incorporated Western technology. Other merchants, they showed, apparently integrated modern managerial practices into their firms, thus altering the social relations of production from feudal clientage systems to manager-worker schemes.[6] These approaches are evidence of a natural indigenous growth of capitalism in China that suggests further success in the years to come. While these studies indicate that future research in China may be very productive, the openly political questions being addressed may continue to hinder the subjective answers they find. More importantly, like the Western studies, Chinese scholars ignore the local concerns of modernization and fail to see the roles of local elites, county magistrates, peasants, and coolie laborers in the negotiations that helped define the nature of the production process.

My book is designed to address these problems and provide a new strategy for understanding industrialization in China. I begin not with a focus on the managers who founded the mines or the unions that organized the workers but with the people who lived in the villages and towns, who worked the rice fields and coal fields, who sold and purchased coal and other products in the local markets for centuries before the industrial firm was established. I examine the county gazetteer, the local document that provides descriptions of local lineage leaders and county officials throughout the Ming and Qing dynasties, and includes entries on cropping patterns and farming strategies, stories of commoners making cloth and marketing their wares, and vignettes on early mining as well as the coming of the German engineer. I supplement this with other documents written by the county magistrate and the Chinese and German mine managers as well as biographies and memoirs of elites and commoners alike. While these documents provide me with the broad brush needed to cover the canvas, the day-to-day decisions and activities were mostly explained in excruciating detail by the hundreds of letters and telegraphs collected in the Sheng Xuanhuai Archives. These letters, which include correspondence to and from Sheng and his associates, were published as raw data in several large volumes that are essential to understanding this history.[7] The correspondence includes vivid descriptions of people on the ground in Pingxiang County complaining of excessively rainy weather that hindered efforts on the ground, tales of corrupt officials who were fired due to insubordination, and depictions of the political friction between managers and local government officials, gentry, and workers alike. Collectively, these documents portray the ongoing struggle and negotiation that created, hindered, and altered the industrialization scheme in the county.

In all, I show that mechanization of the coalmines necessitated the separation from their poorer brethren of the lineage leaders who had secured incomes and safety nets for both strata of the community throughout the last dynastic era. Patron-client relations that allowed landholders to receive rents from farming and mining also assured some fellow lineage members access to subsistence through toiling on gentry properties. Industrial mining, on the other hand, required the break-up of this pattern and the concentration of property in the hands of the modernizers whose capital was far greater than that held by the local lineages.

Furthermore, I argue that within the county the role of the state changed as regional officials and their coterie received the imprimatur of the court to attempt this industrial project. Chinese government evolved from an essentially paternal institution with policies based primarily on hands-off designs into an active managerial administration that sought to transform the local economy. Almost without warning, court leaders ordered the county magistrate and his subordinates to force abrupt and substantial changes in landholding patterns and economic strategies on their districts, even when the inhabitants opposed it citing the Confucian principles that had legitimized the state for centuries. The county magistrate, whose job had usually been to maintain the peace and security of the district, was now called upon to purchase lands from elites and nonelites against their wishes. A mine bureau was founded to reorganize mines and transportation systems and support the implementation of a new economy that was supervised by a foreigner with no background in the values and ideals of the population.

In fact, the foreign technology that was installed in Pingxiang County was itself designed in Western Europe to solve the problems of increasing production with smaller populations and higher paid and skilled workers, a set of issues that were not remotely similar to those being experienced in the densely populated, unskilled, and undeveloped reaches of China. Moreover, German engineers transformed the local working population by forcing them to utilize the technology as it was designed. They brought Western machinery into the mine in the hopes that Chinese workers could be trained to use it properly and efficiently. When workers failed to achieve the level of output the engineers required, the foreigners beat and scolded them for their incompetence. Their methods of brutality and ambivalence incited acts of sabotage and violent resistance by people of the community, forcing the Germans to flee in fear for their lives on several occasions. Events such as these indicate that the problems of modernization in China were not simply due to the inability of the court to properly utilize capital, nor were they due to the relative success or failure of the working classes to create unions and develop into a proletariat. Rather, this study shows that the failure of foreign engineers and others to assimilate all of the county's inhabitants into a united community dedicated to the modernization process hindered the transformation of the economy and society and sparked not an industrial revolution in Pingxiang County but the Communist Revolution.

Because this book examines the actions of several actors or sectors who worked in tandem, in opposition, as well as parallel to one another, the narratives do not always flow in a strict chronological order. Thus, while miners are extracting coal under the leadership of one foreign engineer, land is being purchased for the construction of the railroad by another. Given the conflicting and overlapping timelines, this book works on a loosely defined chronological outline while breaking up essential actions into chapters.

In the first chapter, I examine in essentially classless terms the development and strategies of peasants for subsistence and the role local coalmining played during the late Ming and most of the Qing eras. Using the arguments of Chayanov and others, I show that peasants sought out whatever strategies suited them for the subsistence of their families. In the case of Pingxiang County, easy access to bituminous coal supplemented rice and wheat harvests. Men and boys brought whatever tools they owned and dug shafts up to 100 feet below the mountain surface. I go on to show that, unlike the arguments put forth by Pomeranz, Chinese markets were likely available for coal production that might have sent the mineral throughout the region, including along the Yangzi River. However, the transportation costs and lack of demand for new forms of economic growth hindered regional demand and encouraged local consumption.

In the second chapter, I emphasize the period from the late seventeenth to the late nineteenth century as lineage leaders increasingly saw avenues of financial advancement through the local marketing of coal. Using their patronage powers, lineage leaders took direct control of the coal deposits and rented them out to the peasantry in deals similar to the rental agreements for their rice paddies. While no doubt many of the mines in the county continued to be used as common lands during the late Qing era, the best fields were gradually taken over and brought under lineage control. Among the most, if not the most, important of these lineage leaders were the gentry of the Guangtaifu Lineage Trust headed by the powerful Wen lineage, whose members included not only highly successful literati but also rich men whose control over the mines led to their being termed "mountain lords." The lineage trusts relied upon their control of the mines for continued status and wealth, and in return their rental agreements provided a type of safety net to the men who worked the mines.

In the third chapter, I begin to show the significant changes taking place among the empire's most powerful leaders, focusing on the period from the late 1880s to the mid-1890s. I focus on the Huguang governor-general Zhang Zhidong and the official-merchant Sheng Xuanhuai who established coteries of Chinese merchants, engineers, and officials as well as foreign engineers and investors to create an industrial scheme designed to manufacture railroad track, weapons, and other industrial machinery. This amalgamation of trained and semi-trained staff was sent throughout the empire in search of the best natural resources available for their endeavor. Along with the iron ore mines located in Daye County, Hubei Province, they discovered deposits of high-quality bituminous suitable for smelting iron in Pingxiang County. These men attempted several strategies for securing the mineral, including subcontracting the lineage-led Guangtaifu Lineage Trust to increase productivity to levels needed for a modern ironworks and establishing a mine bureau made up of Zhang and Sheng's appointees. When this bureau determined that the local lineage was simply too entrenched in a premodern world and could not make the needed changes, the men attempted to skirt the lineage leaders and establish a new centralized mining scheme.

Chapter 4 examines the forced purchase of the county's property deemed necessary for a modern mining system. Under the leadership of the first German engineer, Gustav Leinung, the mine bureau put forth a scheme that called for the complete takeover of the county's mining property. Chinese managers were called upon to use eminent domain to purchase by force the mines Leinung required for a fully mechanized mine. Not only were small mines that dotted the fields taken over by the new mine bureau, but the Guangtaifu Lineage Trust was purchased from the Wen lineage almost certainly against the wishes of that powerful family. Moreover, when the decision was made to lay a railroad track from the mine to the river valley in Hunan Province below, the county magistrate was ordered to purchase some of the county's finest rice fields to fulfill the needs of the proscribed route. These stories tell us that as foreign engineers platted out the territories needed for modernization, Chinese managers and government officials complied with their demands to force some of the county's most powerful men to hand over their wealth in the interests of a Westernization scheme.

Chapter 5 once again overlaps chronologically with the previous two chapters as I focus on the actual mechanization process. As soon as Leinung was sent to Pingxiang County in 1896, he and the other members of the mine bureau began to devise plans for integration of Western technology into the county's economy. In a series of letters and conversations, Leinung requested and was provided with machines and tools required for the most effective method of underground mining. To run this complex system of electric machinery and all the supplemental tools and devices required, the bureau recruited a labor force from among the peasants and miners in the region that would leave their families and farms, their lineages and communities, to work year-round in the newly developed mines. While the Germans hoped to maintain direct control of the labor force during their working hours if not also in their social lives, in fact, they returned to patronage relationships through the use of contract labor systems. These contract labor bosses not only undercut the training and oversight of the German engineers, but they were integrated into the local secret societies, illicit communitywide organizations that called for the end of the dynasty and included managers and patrons of opium dens, prostitution halls, and gambling casinos. While the secret societies did not represent a modern form of labor union, they did act as a social group that fought for better benefits and pay, much to the consternation of the German and Chinese mine leaders. Finally, I show that the mines and factories in Pingxiang County in Jiangxi Province and the Wuhan Cities' factories and Daye County mines in Hubei Province were amalgamated together as a modern factory system collectively referred to as the Hanyeping Coal and Iron Company, Incorporated. This new corporation unfortunately failed to reap profits and instead put China into such deep debt that it helped sink the last dynasty. Sheng Xuanhuai, now manager of the entire scheme, fled for his life as the court collapsed and he was blamed for much of its problems.

Even as Sheng's industrialization project was sinking the empire as a whole, a similar, more local, version of this friction between his mangers on one hand and the county's population on the other is the focus of the sixth chapter as I discuss local resistance of both elites and nonelites alike to the imposition of the mine on the county. This chapter begins once again in 1896 as the German engineer Leinung arrived to investigate the county's minerals. The German's presence quickly piqued the interest and anger of several sectors of the county. Even as petty miners planned to jump him and hit him with rocks upon his arrival, students and even local gentry viewed him as a potential threat to their way of life. Gentry leaders, in fact, turned to their officially sanctioned levers of power, writing big-character wall posters and memorials denouncing the modernization plans being contemplated. At least one powerful leader named Xiao Liyan fought tirelessly to keep the mining scheme out, and when he failed at that he taxed their revenues to pay for a school he founded. Similarly, powerful leaders of the Wen lineage, having been stung by the outsiders, sought to take over a local iron ore mine they feared would be the next target of the insatiable foreigners. However, it was the commoners who brought about the most significant resistance against the mine and its leaders. From the early twentieth century, contract labor gangs successfully fought for better working conditions and wages. Their efforts came to a head in 1906 as the Ping-Liu-Li Uprising—one of Sun Yatsen's attempts to overthrow the last dynasty—combined Japanese-educated students from Pingxiang County and local secret societies into a violent rebellion that spread from the mines and throughout the surrounding highlands. The rebellion was not simply a class war, since young gentry influenced by Western ideology joined with peasants in the rebellion while older lineage leaders recruited commoners to fight against their fellow peasants in quickly created militia. Moreover, even when the state crushed the violent uprisings, some members of the local community—elites and nonelites alike—quickly went back to local mining along the lines they had for centuries before, showing once again that modernization attempts failed to fully transform the mind-set of the local community that continued to view their county's minerals as a part of their patron-client system of subsistence agriculture.

The book ends with a conclusion that sums up these chapters. I discuss the implications of the historical events of this book. Specifically, I show that the mechanization of Pingxiang County dramatically increased mineral output and modernized the overall economy. However, the social and political consequences of that policy altered the lives of commoners and elites alike. It pitted local gentry, the state, and the commoners against each other and even caused rifts within each strata. Thus, the Pingxiang coalmining scheme that was part of the late imperial policies called the "self-strengthening movement" ironically weakened and atomized the society and sparked its destruction.

This book provides a study of the tensions and negotiations, successes and pitfalls, of rapid industrialization as it was experienced on the periphery, where natural resources and surplus labor power intersected with the global economy. It shows clearly that to fully understand the history of one of the largest and most modern coalmines in China, we must fully understand the premodern society and economy, its rules and assumptions, its avenues of success and its strategies for survival.

1

Scratching the Dirt, Digging the Rocks

The Economy and Technology of Late Imperial Era Pingxiang County

In order to understand the impact of mechanized mining in Pingxiang County beginning in the late nineteenth century, I begin with a study of the decades just prior to modernization, when coalmining was integrated with family farming and local trade. While peasants' production in this region was not unique, their strategies for using the local natural resources and environment influenced their lives and the history of this county. As the mountainous soils and bitter winters altered the possible farming strategies for peasant families, so too the natural paths of the navigable rivers that flow east and west and dangerous and cumbersome peaks to the north facilitate trade with some towns and cities while hindering communications with others. In this manner, production of crops and coal allowed for local trade in family necessities within the county. However, because of China's lack of investment capital and the peasants' desire to maintain a subsistence-based economy as well as the high costs of transportation of bulk goods, the "take-off" spark that brought about an economic and technological transformation in Europe did not happen in China. Rather, the Chinese chose to maintain strategies that both assured them of subsistence and occupations even at the cost of expanded production.

In this chapter I first describe the world of peasants in this county in the nineteenth century and place that world into a context of both agriculture and mining, family and labor, local consumption and regional markets. After a brief description of agricultural practices, I focus on the place of coalmining in the lives of these people. Then, I examine the markets and strategies they provided for local producers and subsistence needs. Finally,

I examine regional trade and empire-wide interaction where just beyond the confines of Pingxiang County an explosive market in coal was emerging that was whetting the appetites of some Chinese merchants and foreign consumers but not the producers of the Jiangxi provincial highlands.

The Setting: Life in "Duckweed Township" and "Peaceful Spring"

The county of Pingxiang, which can literally be translated as "Duckweed Township," is located in Central China south of the Yangzi River valley and the Wuhan Cities. It is nestled in the Jiangxi provincial highlands, straddling the Luoxiao Mountains, which delineate the border between Hunan and Jiangxi provinces, the two bowl-shaped regions to the east and west. This macroregion, which is an upside-down-shaped "U," features three major rivers: the Xiang River of Hunan to the west; the Gan River of Jiangxi to the east; and to the north, the Yangzi River joins the two together as it collects the runoff of each of those rivers and continues on its eastward trek toward the Pacific Ocean.[1]

The county's topography is mostly rolling hills along the east-west horizontal. Though most of the county stands at an elevation of from 1,500 to 2,000 meters, the highest peaks along the northernmost and southernmost reaches of the county rise to more than 2,000 meters above sea level.[2] Travelers in search of an easy path from Nanchang, the capital of Jiangxi Province, to Changsha, the Hunanese provincial capital—that is, from one side of the upside-down "U" to the other—cut across the less-developed highlands by going through the corridor of Pingxiang County. This trek would include bypassing the county seat of Pingxiang City, among other towns and villages. As was standard for Chinese county seats during the late imperial era, Pingxiang City was enclosed by a brick wall that delineated the city limits. The city's design did not follow standard Chinese practice in that the walls were not properly squared off, as there was some rounding of the southern edges and the entire city was not strictly aligned with the cardinal directions as was deemed proper for a bureaucratic center. It was surrounded on the east, south, and west by the navigable Ping River, which flowed from east to west. To the north stood a large mountain that gently spread into the city walls and provided a backdrop for the magistrate's yamen, partially fulfilling the geomancy requirements of the city's location. Pingxiang City was also a major economic center and was well linked to all the other important markets within the county and beyond. The city contained several market areas, including one located outside the "Minor Western Gate" near an intersection between the Ping River and a bridge that joined the city to several overland routes. From this market a traveler could walk inside the gate and find themselves in the gentry's district, where the most illustrious people in the county lived or spent their time. Many merchants, however, may have traveled from the Pingxiang City market to other markets in Pingxiang County and neighboring counties. Because the many rivers in Pingxiang County all descended from the Luoxiao Mountains into the lowlands on either side, virtually all the rivers and therefore the major trading routes, traveled in east-west orientations.[3]

As travelers left the Pingxiang City market, they could go eastward toward the Jiangxi provincial lowlands. East of the county seat the traveler would quickly run into the town of Anyuan, referred to as "Peaceful Spring" by the American minister Walworth Tyng using the direct translation of the name. Throughout most of the eighteenth and nineteenth century, if not before, Anyuan was a small village with several mining huts and opened mines dotting along the hills. Tyng described the walk between these two locations as a "beautiful one in October or November. The road winds among the hills, which in November are covered with blossoming tea-trees."[4] Not far from there, one could take a boat down the Yuan River to the major market at Luxi and into the Gan River valley of Jiangxi Province. The early twentieth-century county gazetteer, which includes a wealth of entries of stories and lists of facts both small and important, indicates that Luxi allowed for overland and river traffic and that for traveling merchants it was a center of market traffic, "like the spokes on a hub." This market town, which reportedly held several tens of thousands of people in the late Qing dynasty, was an important location for administrative business as well as marketing needs.[5] The traveling guests, particularly those of high rank, could stay at the Luxi Guest House, which was constructed in the late sixteenth century at the end of the Ming dynasty. Other shrines and official buildings described in the gazetteer indicate that Luxi may have existed in the Southern Song dynasty in the thirteenth century and functioned for hundreds of years as a stopping point for officials who traveled from Jiangxi Province into Hunan Province or for those descending from the county into the provincial lowlands.[6]

If travelers left the market in Pingxiang City and proceeded westward, on the other hand, they would descend the Luoxiao Mountains along the Lu River into Hunan Province. When Tyng took this trek from the city of Pingxiang to the Hunanese lowlands in the early twentieth century, he described the route in majestic tones:

> The trip back to Changsha was a glorious one after the October frosts. There is a sort of tree that spots the landscape with vivid red, warm as the send-off of our friends who came to see us off. Between the great hills are terraced fields of golden grain, uncut here and there reduced to stubble. . . . Ping-hsiang is most picturesque of all with river, bridge and wall—from city wall a stretch of plain to the greater ramparts of the hills. The oranges are golden in the autumn, as in the Tuscan orchards, and the pumeloes, show like small yellow moons against dark leaves.[7]

The most important city of Pingxiang County along this route was Xiangdong, a small market town that reportedly contained "more than 400 merchant and commoner families."[8] Like Luxi, Xiangdong contained a Guest House built during the Wanli reign period in the late Ming dynasty. From Xiangdong the traveler would rather quickly enter into Liling County in Hunan Province. The Lu River passes by Liling City, a major marketing and administrative center, and then proceeds in a southwestward direction to Lukou, the city that, as the name implies, is located at the mouth of the Lu River. Here, the river empties into the Xiang River that in turn flows past the major regional rice market city of Xiangtan and then the Hunanese provincial capital city of Changsha. From Changsha, large and small boats followed the river into the Dongting Lake and then from there to the Wuhan Cities and into the Yangzi River to Shanghai and the Pacific Ocean.

Finally, travelers could also leave Pingxiang City and walk along the northern overland footpaths to the important merchant and mining town of Shangli. While the east-west corridor of Pingxiang County was largely sloping toward their respective river valleys, the trek north is much more dramatic. Mountains emerge from the lowlands in sharp and grandiose manner, making the trip northward a difficult one. Once travelers arrived in the northernmost region of the county, they entered Anle Township and the northern market town of Shangli. The Pingxiang County gazetteer lists the population of the city at "300 to 400 merchant and commoner families." Also, the gazetteer does not list a Guest House in Shangli or any buildings as old as those found in Luxi. This indicates that Shangli was a newer city, likely developed after the Ming-Qing transition, and was primarily a market town containing more mining merchants rather than a center dominated by gentry elites.[9] In fact, the area around Shangli was a small mining community that continued to function and prosper at least into the early twentieth century.[10] Since Shangli was difficult to reach from Pingxiang City, it developed much closer cultural and economic ties with the Hunanese counties to its west and provided goods such as coal and tea to the markets immediately outside of the county boundaries more than to those within Pingxiang County.[11]

Family Farming Strategies: Subsistence through Self-Exploitation

In this isolated highland, the mountains provided both minerals for heating and cooking as well as flowing streams that watered the soil needed for farming staples and cash crops. Summer rice agriculture depended on the labor of men, women, and children, who each did their part while fall crops and labor strategies varied from region to region and family to family. Then in the winter months mining was almost certainly the most productive form of labor among peasant men in the county. Hundreds of small mines were opened throughout the county that exposed mineral deposits, including lead and iron ore as well as coal.

Mineral extraction, like handicraft work and farming, relied on the availability of laborers to complete the tasks. The county's production schemes followed the broad outlines theorized by the nineteenth-century peasant scholar Alexander V. Chayanov, who surmised that families sought out family labor strategies that balanced production output with the subsistence demands of their members. Put simply, rice farming strategies required each family member to perform duties following culturally defined gendered divisions of labor and relative skill levels. Subsequently, during the off seasons and off-peak hours of the day, men, women, and children performed tasks that in some way augmented their farming output and added to the overall output of the family's total labor. Chayanov explains that such small tasks were viewed by the family members as "drudgery" and that the payoff was so small as to constitute "self-exploitation," but the members continued to perform these tasks to assure their survival.[12] As each family moved through its natal cycle, it altered the tasks members were able to perform. That is, it developed a production "portfolio" that suited its members' needs.[13] In this manner, peasants within the community, as well as those located in different locations or even those who lived in different times, emerged not as a monolithic class but as various strata of "peasantries" that included some who focused nearly all their laboring energies on farming and others who devoted greater efforts to mining or other nonfarming pursuits.

In Pingxiang County, the people enjoyed a prosperous agricultural economy throughout much of the Qing dynastic era in the eighteenth and nineteenth centuries. In the summer months, rice was the focus of most farming strategies among peasant families, as about 80 percent of all cultivated lands in the Pingxiang County area were irrigated and used for wet rice agriculture.[14] The rice was watered by a combination of the heavy spring and summer rains and hydraulic systems that used water from the rivers that flowed down the mountain slopes. Western travelers who trekked across the Hunanese highlands in the late nineteenth and early twentieth centuries wrote of the ingenious and labor-intensive methods of irrigation that stretched across the fields.[15] Similarly, one Chinese official noted that the systems of dams and wooden derricks and pipes used for agricultural hydraulics in Pingxiang County and the neighboring fields descending into Hunan Province were more complex than any he had ever seen.[16] With the beginning of the fall, some peasants in Pingxiang County farmed a second rice crop. Much more commonly, families harvested winter wheat and hardier crops.

One very colorful section of the 1935 gazetteer written by an unnamed author provides a description of the planting schedule based upon the Chinese calendar and farming proverbs. It explained that the farming schedule begins at the end of the period known as "Excited Insects," or March 5th to 18th. "After the Excited Insects," wrote the author, "you must have cold. This is called 'Freezing Worms.'" Furthermore, the proverb explains,

> "On Qing Ming the frost ends and there is no snow on the land." During the "Grain Rains" (April 20th to May 4th) it should rain. If there is no rain, then it will be difficult for hoeing and plowing. As the proverb says: "If there is no rain during the 'Grain Rains,' then give the land as repayment back to the landlord."[17]

Some of the rice was of the early-ripening variety that could be harvested as early as July, allowing for another crop to be planted and harvested in October. In about half of all lands, a second crop was interplanted with the first, allowing the peasants to use the ample water supplies while avoiding the problems of the growing season that was too short for two complete successive rice crops. The remainder of the rice was allowed to mature throughout the entire summer and harvested in autumn in the month of October.[18]

The late summer months bring the hot temperatures that dry the wet croplands and supply the rice crops with needed sunlight. The gazetteer author wrote that "as the proverb says: 'if it is not hot during the 6th month, then the five grains will not bear fruit." ' This is followed by rains in August during the period known as "The Beginning of Autumn" (August 7th to August 23rd). During this time the peasants believed that if the rains failed to arrive, then the harvest would suffer.[19]

Finally, in the late autumn months the climate becomes cooler and the rains continue to provide for late-ripening crops. The author of the gazetteer entry recounts one proverb that explains, "In the spring if the land is without rain, do not plow the fields, in the fall if there is no rain, do not sow the gardens."[20] Beginning in November, between 20 and 40 percent of the northern half of Pingxiang County was used to harvest winter wheat. Similarly, barley was commonly found in the region in the winter months. Summer and fall crops of corn, sesame, tea, beans, cotton, and other supplemental crops were also noted by Buck in his research.[21] Along with the extensive fir tree forests, which grew both naturally and with active planning by the lumber workers, could be found chestnut trees, from which the town of Shangli takes its name.[22] Animal husbandry and hunting and fishing provided such supplemental proteins as poultry, pork, and fish to augment family farming output.

Not all farming in Pingxiang County was centered on rice agriculture. In the less densely populated southern townships of Pingxiang County where the soils were poor and access to markets was particularly difficult, some peasants gradually turned to sweet potatoes for staple food production.[23] Beginning in the nineteenth century and continuing into the early twentieth century, the American crop spread throughout China and altered the strategies of many peasant communities, including southern Pingxiang County, where nearly 20 percent of the land was used for the farming of the new crop. In some farms in southern Pingxiang County, in fact, rice was not grown at all, but instead peasants farmed sweet potatoes in both the spring and the fall and then harvested in the fall and early winter, respectively. Once harvested, the poorer families who could not afford to secure rice for their diet consumed this new crop as a staple.[24]

The labor required to farm two crops of food and provide fuel for the household involved an elaborate set of strategies for family labor. At the simplest level, Chinese often stated that "men plow, women weave," meaning that men were in charge of most of the physically intensive labor required by farming, while the women tended to do the supplemental tasks of producing textiles and other home-based tasks.[25] However, even this is too simplistic, as women and children performed many tasks in the fields to support men's more muscle-intensive labor. Where rice was grown, for example, women and children pulled up weeds and husks at the end of the season.[26] Sweet potatoes, too, required extensive labor of the women and children as well as the men for a successful harvest. Taken out over decades and centuries, following Chayanov's theories, it is obvious that as young families aged and their members became older, the relative labor and input strategies utilized by each family changed to meet both the available labor inputs and the consumption requirements.[27] The degree to which a given family farmed beyond their own small plot of land, worked in factories or mines, or sought out occupations as maids or servants to the elites in the community changed from one family to the other and from year to year.

So, to supplement agricultural output provided by family labor, peasants turned to a variety of by-employments or nonagricultural labor. Little data exist telling us what sideline industries were found in the county, though many forms of artisanal work and even firecracker production were common. Most likely, many peasants engaged in farming-related activities. For instance, one entry in the gazetteer explained that during the late autumn and winter months some peasants sold cloth to supplement their incomes. Supporting evidence for this contention can be found elsewhere in the gazetteer. In one particular case found in the section on venerable women, a née Peng was said to have arranged for her two sons to marry and trained the daughters-in-law to spin thread for the market while her sons sold some of the grain they harvested. Though née Peng and her family were poor, they engaged in several farming strategies that collectively supplied them with their subsistence. Not only did née Peng's sons provide labor for their own needs, but their marriages also provided added labor power that augmented the family's production as well.[28] Some peasants also engaged in livestock fowl husbandry for sale in the local markets. Among the most significant types of land and waterfowl raised were the ducks peasants tended to, keeping as many as twenty birds at a time, a large number for the region.[29] As will be seen in the next chapter, some members of the county emerged as landholding elites who no doubt turned to some of their poorer brethren to do any number of tasks, including construction, child rearing, and even maid and butler services. In this way, men, women, and children worked for the landlord families in a number of capacities based on their ages, genders, and personal skills.

In all, peasant families turned whatever labor combinations suited them to provide for their subsistence. The variations of staple crop farming, sidelines, and even service jobs allowed them to provide for the subsistence of the family. Moreover, as each family in the community moved through their natal cycles, sought out jobs of various forms, and utilized and altered the environment, jobs that were selected for subsistence by one family for several years might eventually fail to satisfy that family's needs. Those jobs might be, instead, picked up by another family that could not be satisfied by the task at hand previously but would presently find it suitable. Over generations, such a community came to assume that the workable options for production and consumption were available when they needed them. A sense of a bond between the natural environment and the human-made community developed that each individual and each individual family counted on for survival throughout their lifetimes.

Mining Strategies and Technology

In the early twentieth-century gazetteer, a Pingxiang County member proudly proclaimed that the county had "minerals in store (that are as numerous as) the hairs of a pony's tail."[30] In fact, the county holds some of the province's largest fields of gold dust, lead, copper, sulfur, antimony, lime, manganese, and iron as well as porcelain clay.[31] Even still, the deposits of high-grade bituminous coal were by far the most significant and abundant source of family-required resources and dependable by-employment in the county. Geological studies show that beginning in the Paleozoic era, the county's mountains were covered with ocean greenery and dense forests that solidified and transformed into various grades of coal over the millennia.[32]

Over the last two thousand years local inhabitants of the area utilized whatever tools and skills they had at the time to extract coal from the mountains to heat their homes and fuel their kitchens.[33] At least as early as the Han dynasty some inhabitants extracted coal for their daily needs.[34] Moreover, several hundred years later, during the Tang era, local inhabitants of Pingxiang County reportedly opened small mines for extraction and personal use.[35] By the early thirteenth century a local gazetteer and another publication indicated that the county's output was quite exceptional, boasting one of the two best coalfields in Jiangxi Province, in fact.[36]

By the height of the Qing dynastic era in the eighteenth and nineteenth centuries, Pingxiang County peasants developed and employed strategies to utilize the coal for personal needs and to incorporate coalmining into their labor portfolios. Mining was used as another by-employment strategy that was to supplement their farming and was done almost exclusively in the winter months after the last harvest was completed. Beginning in December, the cool and dry winter season begins and lasts for about four months, at which time the temperatures average about 6 to 8 degrees Celsius. These near-freezing temperatures are supplemented by frosts almost every day, and some snow and freezing rains do fall. The precipitation quickly comes and goes and melts, leaving the ground soft and wet. Local proverbs stated:

> One inch (*cun*) of snow turns into one foot (*chi*) of mud
> One foot (*chi*) of snow turns into ten feet (*zhang*) of mud

The gazetteer further explained that during the height of winter the frost takes over and some sleet and freezing rain fall heavily on the houses and make icicles off the eves. In fact, one proverb exclaims, "The winter is a cold and miserable time."[37]

Even as the winter cold was seen by many peasants as "miserable," the local winter conditions made for good mining weather. Since the winters were not excessively cold, the temperature difference between the wells and the land above was not especially dramatic. More importantly, the dry winters protected the miners from the dangers of water damage that can destroy the mines and even kill the miners working below. In those winters when the rains fell harder and longer than usual, the damage to the mines—not to mention the homes and fields—was devastating. The excessively wet winters damaged or destroyed the mines and slowed the transportation of the mineral to the homes and markets in the valleys below.

During these brutal winter months, the women and small children of the family turned to indoor activities like cotton spinning or handicrafts and the men and older boys grabbed whatever tools and supplies they could carry and they left the family farms for the mines in the mountains. When they arrived in the mountains, they sought out the best locations for coal they could find. That is, they searched for the best coal located as close to the surface as possible. In many cases the men turned to pits they had used the previous year and simply began yet again. However, due to the changing seasons and weather patterns and general neglect, these shafts were usually destroyed and the walls of the shaft crumbled in, making them too dangerous to work. So instead, they simply planted a flag in a new location nearby claiming the land for themselves and started excavating a new mine. In this manner, flags and dilapidated mines were strewn throughout the highlands providing the next generation of miners tangible evidence of past successes and failures.[38]

Even as the abandoned mines were potentially dangerous traps that many unsuspecting people fell upon, the new mines were equally dangerous and often crudely constructed. Men and boys who worked them did so knowing that the chance of injury, if not death, was high. To enter the mine, the worker walked down a crude bamboo ladder until he reached the mineral seam. In many cases, the miner continued downward to a second and even third seam in the mountain. Most of the coalfields contained two seams located close to the surface that produced poor-grade bituminous coal that was lumpy and therefore easy to transport and use in family furnaces. Moreover, the coal was low in ash and so would produce little smoke when fired.[39] Nearly all the small pits in the Anyuan area were dug down to retrieve the coal from these two seams. Then, when the mineral deposits near the mine shaft were depleted, miners usually started a new mine to extract elsewhere. However, some of the mines were dug deeper to extract the higher-grade bituminous coal. This coal was softer and lower in phosphorous and sulfur and, though it was high in ash, it was still a better-quality coal than the top-layer deposits. In any case, the deeper the miners descended, the more dangerous the task. And once they arrived at the spot they intended to work, the seams were often narrow and undulating, forcing the worker to move about under the rock following the natural contours of the deposit. For example, one Western observer in Pingxiang County explained that as the miner burrows into the ground, he

> avoids rock so far as is possible. . . . The diggings are largely in the seams and consequently have many torturous and narrow passages. The shaft of the native mine follows the vein from the surface, usually at an inclination of from 20 to 60 degrees. After a varying distance the shaft or drift becomes horizontal and then rises still following the vein.[40]

Miners were required in this manner to descend into the rock and then crawl along the seams picking and digging their way to find the minerals they needed. Most mines were reportedly dug no more than 100 to 150 feet deep.[41] And even at that depth, the German engineer Baron von Richthofen explained, "[I]t proved to be very unsafe," though some wells reached even greater depths. These shafts were only fully excavated at great danger to the workers.[42]

Most of the mines were completed using simple farming tools the peasants brought from home. Being too poor to invest in specialized devices for mining, peasants from the most ancient periods until the early twentieth century employed whatever gear they could bring from home. Since mining obviously began with the simple digging into the ground, one would assume that shovels would be among the most important tools needed. However, even though irrigation agriculture itself required constant attention to dredging silted canals, the Chinese did not develop a shovel as Europeans did.[43] Instead the tools peasants brought to the mines were primarily digging sticks or sharp metal objects like harrows or sickles used to maintain their family's rice fields; the hammers, mallets, and other woodworking tools needed for home or farming construction; some stone quarrying tools for building dams and bridges; and whatever candles and baskets they used at home.[44] Among the only specialized tools most miners used was an iron gad, a wedgelike tool that may have a wooden handle. The method of mining with a gad and mallet is hundreds if not thousands of years old, and the tool designs have changed very little in that time. While many hammers they owned had long handles for leverage, the ones used in the mines were often quite short so that the miner could fit inside the cramped space without disturbing the fragile walls. Unfortunately, this need to utilize short-handled equipment meant that the force they could bear on the rock was diminished, but in many cases the shaft sizes dictated the tool selection. Pick axes and other tools used in the 1920s in local Jiangxi provincial mines had handles as long as two feet. Other long-handled equipment was said to be too long for work inside the shafts but was incorporated into above-ground activities. The local museum in Pingxiang County holds several tools that were used in the premodern mines. Simple axes with medium-length wooden handles predominate the collection and provide a picture of the daily tasks and methods of extraction.[45]

Once the older, more skilled workers extracted the mineral from mine walls, young men or boys brought the coal to the surface by climbing up the bamboo ladder with heavy baskets or carrying poles balanced on their shoulders. This labor—both the transporting of heavy loads upwards of 100 feet to the surface and the work at the surface—was in fact so arduous that it often constituted upwards of 60 percent or more of the total labor power and number of laborers in the mine team.[46] The relative strength of the young boys and the size of the shaft that provided the opening for both carrier and basket constrained the size and weight of the baskets and their contents. For the larger mines, windlasses were constructed over the opening that sent baskets up and down the shaft to the mine bottom. The handheld crank could require eight to ten men to operate in order to bring the heavy mineral to the surface. In some cases, in fact, the windlass was also used as a crude elevator for workers who stood in the basket on their way into the mine. The hundred or more pounds of either a miner or a load of rock placed a strain on the muscle power of the mine crew and in this manner altered the extraction and purification strategies from one mine to another.[47] For instance, in places where vertical shafts had to be excavated so deep as to hinder output, some miners excavated horizontal adits instead. These mines cut into the mountain from the face and directly exposed a single seam. Not only did this method at times improve the chances of extracting lower and more lucrative seams, but it allowed the workers to walk in and out of the adit without the use of ladders or windlasses. This strategy was rarely if ever used in Pingxiang County prior to mechanization.[48]

To prop up the mine shafts, miners used timber along the ceilings and walls. Incorporation of timbers into the mine shaft had to be done properly in order to maintain structural integrity inside the artificial cave. Miners quickly found that timbers should be placed at slight angles when they prop up their shafts. In many cases, miners constructed box-like structures within the shaft to increase the strength and security of the support beams. They also found that in damp shafts timbers became damp and weak. In locations where only poor timbers were abundant, miners had to replace the structures regularly or gamble with their lives as the wood lost its rigidity. In some mines, workers left strategic areas of the rock in place to prop up the ceiling above. This method, referred to as a "room and pillar" technique, was used extensively in some areas of China and little if at all in others.[49]

As has been suggested earlier, excavating wells of one hundred feet or more, especially when they are exposed to the seasonal rains, meant that the wells filled up with water and threatened the efficacy of the mines as well as the health of the men who worked in the deep shafts and tunnels. When there were problems with water seepage, the miners used methods of extraction that were crude but sufficient for the needs of the miners. Throughout China miners who experienced this problem simply opened a new shaft and began yet again. More often, as the rains or groundwater seeped into the shafts, boys and other surplus laborers hauled baskets lined with oiled cloth or leather to bring water out of the deep shafts.[50] Chinese miners are somewhat unique in employing this strategy in that they did not invent the water pump as the miners in Europe did. Even though it seems perverse that the Chinese did not develop this technology given the threat of water damage and injury due to seepage, some Western scholars suggest that Chinese miners did not develop a pump in part because they chose not to. In his study for the *Science and Civilization in China* series, Peter J. Golas informs us that the pump that was used by Europeans as early as the Roman era could have been a useful device had the Chinese developed it sometime during the two millennia prior to the late nineteenth-century Western technological invasion. In fact, he tells us, when Chinese miners did see diagrams of European pumps, they chose not to incorporate them into their production schemes even though the pump designs required little capital investment and fell within the technological levels of the society.[51] Even more likely, Chinese miners chose not to develop or, failing that, incorporate the pump because it did not fit the perceived needs of the mine culture. Given the cheap labor costs of mining in China, technology that Westerners deemed as useful was often viewed by the Chinese as unnecessary. Even more importantly, the Chinese viewed technology that cut labor out of a production scheme as harmful because the same families that consumed the coal also produced the coal. In this manner, petty producers sought to keep their employees working rather than searching for strategies that could eliminate them from the labor force.

It is also true that the Chinese did not simply allow water to ruin their mines. Instead, miners used a number of alternative approaches to alleviate this problem. Following the demands of the miners to keep their labor force intact, these strategies tended to be labor-intensive rather than labor-saving schemes. One method of extracting water that actually does follow the general principles of the pump was the incorporation of bamboo as a natural tube, as described by one Westerner who observed mining in Pingxiang County and elsewhere in the early twentieth century:

> Pumping is effected by manpower, as machinery is never used. A long section of a large bamboo, 6 to 8 inches in diameter, is cleaned out, making a circular smooth pipe. Into one end of this a crude valve is fitted and into the opposite end is introduced a piston with a valve. This pump is laid along the slanting floor of the shaft and operated by a coolie who sits at its upper extremity. The water is caught in a small pool lined with clay from which it is pumped by a second similar apparatus at a higher level. A sufficient number of these relay bamboo pumps are provided to reach the surface.[52]

Similarly, some miners incorporated the commonly used water wheels that farmers designed centuries ago to bring water from one rice field to another. The oblong wheels contained a series of leather buckets or other clothlike baskets to fill with water at the bottom of the wheel's run and then deposit the water when it reached the top. These devices, too, were pedaled by one or more laborers often for hours at a time.

Chinese coalmines not only suffered from too much water but also contained dangerous gasses that came off from the newly struck coal rock and in some cases were so deep that workers nearly suffocated. In fact, in one study of Chinese mineral economy, the global historian Kenneth Pomeranz argues in part that Chinese miners rarely had problems with water seepage but instead had greater problems with ventilation of gasses.[53] His argument is an important one in that it also points to the need by many miners to seek solutions to ventilation. It is true that miners nearly suffocated where the deepest shafts were nearly devoid of breathable oxygen and the gasses that came from some of the coal in Pingxiang County's mines was particularly heavy and dangerous. To alleviate these problems these men sought strategies such as opening a second ventilation well, and at times some miners kept a fire lit in the bottom of the well that moved oxygen through the shafts. More importantly, however, Pomeranz's depiction of Chinese mines being exceptionally dry is quite odd and counter to the findings of most scholars who have studied mining in China. These scholars are nearly uniform in stressing the point that water seepage was a constant and real danger.[54] It is true that both ventilation and water seepage were serious problems for miners. It is also true that water seepage was especially troublesome and dangerous and was handled by local miners in manners that suited their needs and the tools and capital available to them. That is, Chinese miners did not invent the water pump because they did not perceive a need for one. They did, however, find ways of alleviating the dangers of both poisonous air and flooding waters but did so in ways quite different from those employed in Europe.

After coal was extracted from the mines and brought to the surface beyond the dangers of the pits, a number of boys purified the coal. To this end first the boys separated the high-grade coal from poor-quality mineral and shale. Once the best coal was separated from the rest, men coked the coal to produce the high-grade fuel source. To coke coal—a process similar to turning wood into charcoal—one needed to cook the extracted mineral in a controlled furnace until the impurities were burned out of the rock. Miners completed this task by making crude coking ovens. After making long, shallow trenches in the ground, the men filled the ditches with coal and placed brick and clay over the mineral. They then lit the coal and allowed it to burn at a low temperature for two to three days. This coking process separated the pure carbon from most of the volatile constituents and impurities in the raw mineral and made it more effective when smelting iron ores or other minerals.[55] Coked coal's burning temperature was lower than the much harder and purer anthracite. However, when it was fired it put out a constant heat with little smoke that allowed for controlled smelting. This, in turn, improved the quality of the iron produced.[56] Moreover, volatility of the coal was greatly reduced and therefore it could be left unburned longer. And, because the coked coal was pure, it could be transported more cheaply than if the coal was sent with shale and other heavy impurities.

Peasant men completed these tasks in many parts of the county, searching for the locations close to home that held sufficient sources of mineral that could assure them of success in their efforts. For the most part, the coalmines within the county limits predominated in three locations of the Luoxiao Mountains that formed the Jiangxi-Hunan provincial border. To the far north, small mines around Shangli merged into Liuyang County, Hunan. To the northeast, the mine fields referred to as Gaokeng stood near the border with Yichun County. And the center of the county, near the county seat, contained the rich fields around the village of Anyuan.[57] As the county grew into communities centered around one particular market or another or with one lineage or another, the mining topography developed accordingly. Similarly, as certain locations were proven to be more lucrative than others, families altered their labor portfolios accordingly, and in this way their agency altered the history of the county.

I have argued up to this point that the climate, topography, and mineral deposits in and around Pingxiang County were integrated into the annual production and consumption patterns of the local population. The warm, rainy growing seasons and mountain-fed streams provided the conditions for extensive wet rice production throughout much of the county. Conversely, I have shown that the families that settled in Pingxiang County were influenced by their internal gendered composition and natal cycles and were further integrated into a community whose actions were focused around the seasonal clock. In order for families to survive in the isolated highlands, they had to spend most of their year working in irrigated fields of rice while supplementing their incomes with by-employments including coalmining accomplished by the men and boys in each family.

While some families simply gained sufficient benefits through much of their labor, others turned to the marketplace to gain cash and trade value. For them, family strategies that included production of surplus foods and extracted coal could not sustain the family if the surpluses could not be sold in an active marketplace. To this end, local and regional markets established along river routes and natural crossings purchased many of the goods produced by Pingxiang County residents, as traveling merchants supplied them with other goods and services.

Local Markets and the Furtherance of Subsistence

All families needed to engage in some trade to secure certain goods and services they could not provide for themselves. To this end, they produced surpluses in goods they could trade in the marketplace. Marketing strategies were influenced by the relative demand for the product and the cost of transporting that product from the site of production to the markets. Since most staples are both heavy and relatively abundant throughout both Hunan and Jiangxi provinces, trade in these goods was almost certainly contained geographically within local markets. Tea, on the other hand, was light and in higher demand for markets farther from the source. To reach regional markets, consumers and producers alike trekked over narrow footpaths or seasonal river flows to reach the markets. Since many paths into town were slow if not treacherous, they hindered transport methods and schedules and therefore altered marketing schemes further. From this, nested markets of various sizes and specialties arose both within the county and in the regions beyond the county's borders.

Among the goods most prevalent in the Jiangxi provincial highlands was good-quality bituminous coal. Though most peasants did not mine coal primarily to sell, certainly some of the mineral was traded to other peasant families and migrant peddlers who attended the market days. For many part-time miners, the price they received for the coal was miniscule as the transportation cost alone likely priced the mineral beyond the budgets most could afford. Even the cost of transporting coal more than a day's walk from the mine pit was prohibitive for most consumers.[58] For example, the cost of anthracite coal at the mine pit in Leiyang, Hunan Province, was about 2 *yuan* per ton.[59] However, the economist Thomas Rawski determines that the cost of a single ton transported just one kilometer on the backs of workers from the mine shaft to the dock would cost an additional .2 to .35 *yuan* or an additional 10 percent haulage cost.[60] In this manner, miners simply brought as much coal back from the mines as they could and sold it for whatever they could get and considered it an additional revenue for their families. It is possible that isolated markets and high transportation costs were beneficial to miners as it tended to cut off investors who might wish to undercut them with coal from outside the immediate marketing area.[61] However, at the same time, the lack of access to markets meant lower profits or potential profits for small investors. These two realities assured the markets of small amounts of coal demanded for daily use but very little profit required to bring about expansive or sustained output for the county's inhabitants.

These calculations also suggest that at least some Chinese sought to create a year-round mining firm that could provide fuel to the richer families if not to peasants located closest to the mine shafts. To the degree that coalmining was an occupation that paid a wage it is likely workers who agreed to mine for pay earned a fraction more than a farmer, but even this supposition is difficult if not impossible to prove. Figures for Pingxiang County miners' wages are not available, and deciphering an exact conversion rate may be difficult as the values of Chinese coinage varied over time and from locality to locality. Many studies have shown the wages of one occupation or another in a specific location or time though these figures are more suggestive than conclusive. Yet based on these admittedly incomplete numbers, it is safe to assume that the poorest professional miners were making little more than minimum subsistence wages. These wages could not entice men to leave their homes and families to work as professional miners. Those who did attempt to take mining as an occupation were primarily landless laborers who had no family or lineage ties they could turn to for alternative subsistence. More importantly, the poor financial gains promised these men curbed the chance of a possible breakthrough in the Chinese mining sector.[62]

While mining in the late Ming and early to mid-Qing eras was primarily meant for subsistence, there is some evidence to suggest that as the population of the county increased by the mid to late nineteenth century this sparked more trade in the mineral. To this end, individual families joined into larger cooperatives and worked together to excavate deeper shafts.[63] By the height of the Qing dynastic era, new markets were formed and expanded transportation routes and merchants' schedules were established to supply the developing local economy.[64] The concomitant increase in total demand sparked expansion of mining among the seasonal workers. In some mines, the shafts became increasingly deep and the labor extreme. One observer of Chinese mines wrote that he had descended into shafts that were 600 to 800 feet deep. He indicated both amazement and anxiety at the effort of these miners when he wrote, "In most cases the wonder is not that the miners have stopped where they did but that they were able to penetrate so far and won so little from their efforts."[65] From this we can see that workers were indeed willing to engage in difficult and dangerous work for the addition of miniscule added wealth to the family. The local markets provided these men with locations to trade their goods and take home cash or exchange items needed for their family's survival. In this manner, the marketplace was an essential feature of their lives and was well integrated with their farming and mining labor. However, because the costs of transporting the coal in any great distance dramatically increased the cost of the good, the low capital demand for cash crops and commodities failed to spark the evolution of subsistence mining and sales into the kind of capitalist markets that were developing in Europe at the same time. Thus, the "take-off" phase did not materialize.

Local Coal Production and Regional and International Markets

Because the Chinese economy was labor rich and capital poor this affected the farming and mining strategies the people employed, and it had a significant impact on the nature of the markets. Local trading centers and market towns were designed to provide for the subsistence of the people, and in this they succeeded. At the same time, in the mid to late nineteenth century, regional markets and international trade lagged in China compared to contemporary markets in Europe. Chinese populations sought production strategies that promised jobs and the greatest chance of subsistence for their members and viewed the promise of more efficient production schemes or more elaborate marketing strategies as potentially dangerous to those goals and therefore counter to these plans.

In Pingxiang County, coalmining as well as rice farming and cash cropping provided in theory a chance to engage in transformative regional trade, though that opportunity never arrived. Larger markets located along the greater waterways outside the county might seem to a modern capitalist as an opportunity for economic advancement. However, trade in high-grade bituminous coal did not take off as a regional commodity. Indigenous development of coal sellers and buyers did not lead to new strategies for trade or transportation or to avenues of investment or mass marketing. Rather, even in the nearby lucrative markets of Hunan Province and in the international market city of Shanghai, the demand for coal did not reach the mines of Pingxiang County. Rather, both markets remained content with less costly items that employed labor-intensive rather than capital-intensive production. And in this way, the indigenous economy did not entice greater investment or the search for new forms of marketing of their coal into the larger markets.

The most likely destination of extended trade from Pingxiang County was the Hunanese provincial lowland cities to the west of the county. Local producers who could find a viable strategy for transporting their goods nearly 100 miles down the county's rivers to Hunan Province's market towns or more than 200 miles further down the Xiang River to the Wuhan Cities in Hubei Province could have access to hundreds of thousands if not more than one million potential buyers in each marketing city.[66] And to this end, tea merchants frequently used this route to sell their product in the Xiang River valley markets. These men who were part of the Liling-Pingxiang Boat Guild transported some goods, presumably tea, to the Wuhan Cities, porting their boats at a special pier built by the guild.[67]

While tea farmers had some success selling their product in Hunan and Hubei provinces, coal miners did not. Even local porcelain producers in such important sites as Liling County and Changsha turned away from regional coal for fueling their kilns.[68] Many potters in China consumed coal rather than wood to fuel their kilns because the heat provided was as high as and attained more quickly than wood. For this reason, the availability of Pingxiang County's bituminous provided a potential alternative to felling the forests for porcelain manufacture. However, even the demand for coal by these potters remained low throughout China as the use of coal had several pitfalls as well. In particular, they complained that coal was too high in ash for making good-quality porcelain, and coal tended to burn for less time than timber. More importantly, to use coal they had to retool their kilns, thus forcing them to rely heavily on this inferior product.[69] Under these circumstances, even in some of the most prosperous factories of the nineteenth century, the demand for coal was insufficient to alter Pingxiang County miner merchants' marketing strategies.

The county's coal also suffered from high transportation costs due to inadequate and seasonal river passages. In particular, the Lu River was so low as to be unusable for several months during the dry winter period. Moreover, the wet months were a time when peasants diverted river water in order to provide for their irrigation systems that watered their fields. Lastly, the Lu River was ill-suited to trade with the most lucrative markets of northern Hunan and Hubei because the river route took a decided turn southward after leaving the county seat of Liling City.[70]

Even as these institutional and natural hindrances to Hunanese provincial markets stifled trade between Pingxiang County and the towns to the west, it was certainly the case that the entire Xiang River basin was a thriving market region in the late Qing era where boats loaded with coal, tea, and rice traveled along the route. We know this in large part because beginning in the late nineteenth century Western merchants and engineers traversed along the Xiang River looking for natural resources. Local Chinese people and government agents alike did not allow Westerners entry into the region until after the Opium Wars, and unequal treaties forced the court to relinquish access to foreigners. Beginning in 1860, under these agreements, foreign steamboats passed inland along the Yangzi River. This opening encouraged a series of foreigners to enter into Hunan Province in search of rich coal deposits.[71] Traveling in Hunan Province was a very difficult matter as the xenophobic traditions of the Hunanese was known even among the newly arriving outsiders. As the American William Barclay Parsons writes, "[E]ven Chinese other than those of Hu-nanese origin have not been welcome."[72] These fears were not enough to keep Westerners out, however, as the desire for cheap fuel and the sense of cultural superiority apparently pushed the travelers into the smallest markets and mining communities in the region. A man named Mr. Dickson in 1861 was the first Westerner to enter Hunan Province in his trek from Guangzhou Province north over the highlands in southern Hunan. From there he traversed the entire length of the province by boat out of fear of reprisals from the local inhabitants. Subsequent attempts by Westerners to enter Hunan Province were at times stifled if not dangerous. In 1865, an American geologist named Mr. Pumpelly attempted to enter Hunan Province from the Wuhan Cities in Hubei Province to the north. When he arrived in the provincial capital city of Changsha, local authorities forced him to return to Hubei. Then, in 1869, the Shanghai Chamber of Commerce hired the German geologist Baron von Richthofen to examine the coalfields in Hunan as he had already done in much of northeastern China. Though Changsha officials also blocked him from entering the city, he was able to examine the other markets and fields along the Xiang River mostly under protection from a Western gunboat.[73] Then when another traveler, one G. James Morrison, attempted to traverse Hunan Province entirely on foot in 1878, a mob came upon him and beat him and so he immediately hired a boat for the remainder of the trek.[74] Finally, around 1900 Parsons arrived in Hunan from the Wuhan

Cities with a retinue of Chinese officials and merchants. Their collective gravitas pushed the officials to grant him permission to enter the city of Changsha as the first Westerner to do so.[75]

These early Westerners' tours into Hunan Province were almost entirely for the purpose of determining the feasibility of purchasing coal in the markets along the river ways. And, to their good fortune, they proclaimed that they found plenty of good-quality coal in Hunan Province along rivers deep enough for large shipping passage. Their published findings make clear that coalmines located along or near the Xiang River or its tributaries marketed their coal dating at least back to the mid-nineteenth century. Based on observations of junk traffic in the winter months when he did his investigation, Richthofen concluded that at least 150,000 tons were being sent by small craft every year. Nine years later, Morrison explained that his observations of boat traffic in almost the exact same time of the year indicated that trade in coal was increasing primarily due to indigenous demand rather than the increased appearance of foreigners and Western river ships.[76]

The junk trade that these men observed indicated that securing coal and other minerals was possible, though several problems were apparent. Richthofen was very impressed with the cost of local boat transportation. Chinese families who established themselves as transportation workers were able to move coal along the Xiang River for less than Western steamships likely could.[77] This was especially important, he felt, because he deemed the Xiang River to be too shallow for large steamships once they reached even a few miles south of the Dongting Lake.[78] Subsequent investigations indicated that steamship traffic was possible at least to the city of Xiangtan and at least during the wet months of the summer. However, in the winter months the water levels were so low that large river traffic could be greatly hampered.[79]

Traffic along the Xiang River into the economic center of Hunan Province posed a problem, however, because investigations found that it was the coal in the southernmost reaches of the province that were best suited for their needs. Nearly every geologist and engineer who traveled along the Xiang River north of Leiyang County mentioned that much of the coal they found was medium- or poor-grade soft bituminous. Since this coal is softer and less pure, and burns at a lower temperature than anthracite, it was mostly being coked and used to fire the breweries, papermaking factories, as well as the petty antimony mines and ironworks factories along the river.[80] However, in the southeastern Hunanese county of Leiyang, Richthofen exclaims that the anthracite deposits were as rich and vast as those found in Pennsylvania. "The whole of south-eastern Hunan may not unjustly be called one great coal-field," he argued. "It is, without exception, the greatest coal-field that has hither come under my observation in China."[81]

Along with stories of plentiful coal and Hunanese obstructions, some investigators mentioned alternative solutions to securing coal that further suggested the prevalence of an even larger indigenous market in minerals and other goods. Parsons notes, for example, that merchants in Xiangtan purchased coal from Hunanese coalmines and then subsequently sold the mineral in markets in Hubei Province and elsewhere.[82] He also found that in the Western encampment of Canton, or Guangzhou, coal shipments from Vietnam and even Japan provided the city with its fuel needs as the transportation costs were lower than shipments from southern Hunan.[83]

If Guangzhou was able to secure coal through international trade, Shanghai promised even more profits for potential traders and mineral merchants. In his discussion of the economic forces in early modern global history, Kenneth Pomeranz argues that Shanghai—or more generally, the Yangzi delta—was incapable of securing sufficient coal reserves to spark an industrial revolution in the manner similar to England because there were no reserves nearby. Unlike the coalfields of England that took advantage of the well-developed waterways linking the northern coalfields to the markets in and around London, he explains, the great Kaiping coalmine was located far away from the commercial centers and well-traveled river routes of the south.[84] Pomeranz's argument assumes two important points that are both provably false. First, he argues that Shanghai was incapable of securing Chinese coal because the only adequate sources were far away. Specifically, Pomeranz notes that China's southernmost provinces contain a mere 1.8 percent of China's total available coal deposits, while its eastern provinces hold only 8 percent of the country's mineral.[85] He suggests that this amount was insufficient to spark industrial expansion. But is he right? According to Golas, China's total estimated coal reserves in the mid-1930s to the mid-1940s was approximately 283,536,000,000 tons. Of that number 7,884,000,000 tons was poor-quality lignite and the remainder was either anthracite or bituminous coal. If we multiply the good-quality coal deposits by 1.8 percent, representing the southern provinces, we are left with an approximate holding of 5,103,648,000 tons. Multiplying it by 8 percent representing the eastern province we achieve the total of 22,682,880,000 tons. Using Golas's figures more directly he shows that Jiangsu Province located directly to the north of the Yangzi delta held approximately 217,000,000 tons and Zhejiang to the south contained 100,000,0000 tons while Anhui, located slightly west of the delta, held as much as 360,000,000 tons of various qualities of coal.[86] In simple numbers, the total available coal in the three provinces surrounding the Yangzi region was approximately 677,000,000 tons. Of that total, over 100,000,000 tons was high-quality anthracite, and nearly all the rest was bituminous, leaving less than 10 percent as poor-quality lignite.[87] While studies such as that done by Pomeranz may focus heavily on the few large mines that were opened in the region during the late nineteenth century, the reality is that even as late as the 1930s, small mines located throughout the empire constituted about one-fourth of the total coal output in China and might have supplied Yangzi markets had there been sufficient demand.[88] In fact, during the late imperial dynastic period, China's total annual coal output was around 500,000 tons.[89] Moreover, given the previous discussion of the mines in Pingxiang County alone, it is obvious that this figure could have gone higher if the market demand had been sufficient.

Second, Pomeranz assumes that Shanghai enjoyed no alternative sources of fuel besides the nearby provinces and mining towns of the Jiangnan region. In fact, Shanghai imported coal from Taiwan and Japan as well as some European countries relying on ocean traffic over river routes.[90] Shanghai's imports totaled more than 110,000 tons of coal annually in the early 1860s, an amount roughly equal to that estimated by Richthofen for Hunan Province. Even more, by the mid-1880s Shanghai's foreign imports of coal rose to more than 380,000 tons.[91] These figures certainly do not point to a market devoid of the products required for their needs, nor do they suggest an economy devoid of expansionist capabilities.

In fact, Pomeranz never provides a sense of what amount of coal would have been required to spark a transformation. Yet it is obvious that billions of tons of good-quality coal were available in mineral fields within the Jiangnan region. Moreover, hundreds of thousands of tons were being produced locally, while several times that amount was being imported from other countries in Asia. Even if this was not sufficient for the complete transformation Pomeranz envisioned, Chinese merchants and engineers might have believed that it was and started the Jiangnan markets along this path. In reality, this did not happen because, as the historian Philip C. C. Huang argued, China's economy was designed to provide for the subsistence needs of the people without turning to the technological changes that push unneeded populations to their deaths. As rising demographic levels required increased production, Chinese peasants did not seek new and potentially disruptive strategies to increase their output. Instead, peasants worked harder in their fields, employing the strategies they knew worked to improve their rice harvests while they also spent even more efforts on by-employments and other nonagricultural activities to provide for their family's needs. The consequences of these strategies for peasant families were that they added to the "self-exploitation" that Chayanov described. For the empire as a whole, they blocked possible solutions to productivity needs, including technological changes along the lines of the Industrial Revolution in England. To the degree that the Chinese needed coal for heating their homes and petty artisanal production, hundreds of thousands of peasants extracted around 12 million metric tons of coal annually and sold it in the marketplaces.[92] Under this system, consumption and production needs were both provided for and in this manner brought about an "economic equilibrium." On the other hand, Pomeranz's assumption that easier access to coal would bring about a spark to the economy assumes that China could sustain a dramatic increase in production that emphasized labor-saving and capital-spending schemes that were counter to China's needs. While admittedly the data is insufficient to resolve this issue with absolute certainty, there appears to be nothing that might support this thesis.[93]

In all, Chinese markets in bulk goods such as coal were plentiful but not transformative. Western investigators who traveled throughout the empire believed that all the materials were present to penetrate into the empire's belly and transform the land into a county that supported modern industrial if not capitalist development. However, to imagine that China was rich in natural resources but ignorant of modern development is to miss an important point regarding Chinese social and economic development. For most Chinese, the markets did not fail to perform as they were needed because they were primarily used for the daily needs and subsistence strategies of the population, a service they performed adequately for centuries. In fact, Chinese commoners were more likely to fear merchants and local elites turning away from their daily needs and moving into new avenues of investments in order to make larger profits for themselves. This fear was bound into the subsistence and labor portfolio strategies that were the focus of their family's well-beings. Therefore, the Chinese did not make the changes Westerners called for not because they did not understand the changes but because they understood the social consequences of these changes and resolved to avoid them at all costs.

This chapter has explained that coalmining in Pingxiang County in most of the Qing dynasty was centered on small peasant families who integrated coalmining into their peasant-based labor portfolio. In order to survive, families had to balance their consumption needs with their production parameters. To this end many in Pingxiang County incorporated coalmining with the farming of staples. While mining was part of the subsistence strategy, the expanding markets in the county sparked some growth in production and sales in the markets. Yet the costs of transportation continued to stifle hopes that more efficient forms of production could spark the levels of investment needed for a qualitative change in mining techniques. At the same time, in the Hunanese provincial lowlands a much more lucrative market in coal and other goods was emerging in the market cities of the Xiang River valley. While Pingxiang County's coal was too expensive to ship to these markets, those economic centers were not unknown to the miners of the mountains to the east. Even though expansion of production and sales to Hunanese markets were theoretically possible, the indigenous economy did not grow to a point that a qualitative change transformed regional marketing strategies, and this left Pingxiang County's farming and coal economy largely isolated from the neighboring cities and thoroughfares.

At the same time, as will be shown in chapter 2, beginning around the last century of the Qing era county elites with access to capital likely decided that investment in the county's mines could become profitable. To this end, they used their political offices to accumulate wealth and their leadership in the lineages to gain legitimacy over the peasantry. What came of this were locally based firms that controlled mining and mining labor throughout much of the county. While these firms were much larger and hierarchical than the family mines, they were rarely more sophisticated technologically. Moreover, men were just as likely to work in the mines as by-employment as they had before, and the coal was just as likely to be used for personal and family needs. Thus, coalmining was integrated into the landlord-tenant schemes that predominated in the rice fields without any significant alteration of technological sophistication or concomitant increase in overall productivity.

2

Relatives, Clansmen, and Neighbors

Local Politics on the Eve of Mechanization

Coalmining and the economy of Pingxiang County functioned within a social and political framework that pitted landed elites and court officials against and with commoner peasants and miners. As the population of the county grew, lineage leaders and landlords increasingly viewed the county's mines as a new avenue of wealth. The county's most powerful men gobbled up coalfields and other mineral-rich land that peasants and other commoners mined to provide for their subsistence, and they then rented this land back to the peasants-cum-miners as commoditized property. This increased the economic pressure on subsistence farmers, coercing them into paying new rents for mineral extraction while strengthening the patron-client bonds the lineage leaders used to control the poorer members of the county. As economic, political, and population pressures grew in the county, the commoner population sometimes turned to violent and nonviolent forms of resistance and noncompliance. However, as the social sectors fought for better conditions of economic and social status, these conflicts did not create revolutionary change. Even when peace brought population increases or, alternately, war brought population decline and death, generations-old social systems still determined strategies of production and tenancy rates.

To fully understand how coalmining transformed Pingxiang County, we must first understand the community's social relationships and the pressures that placed their imprint on the various actors.[1] The modernization of China paralleled in some ways events seen in Europe.[2] Unlike Britain, where the strong state and powerful landholding classes forced the peasants into the markets, the weak political and economic elites in France could not force the peasantry to give up their control over the factors of production and access to survival. Under these conditions, peasants were able to produce for their subsistence while still paying the rents and taxes elite classes extracted from them. This relationship allowed the producing classes—the peasantry—to continue to survive without needing to turn to the market, increased productivity, or modern technology. The elite classes—landlords and the state—failed to gain control of the means of production but were also assured of rental incomes sufficient for their families' reproduction. Neither the elites nor the nonelites could be coerced to change their strategies employing the market or technological improvements. Thus, the system was self-sustaining but it also stifled growth or innovation.[3] Moreover, even as external forces increased demand for the country's manufactures, the transformation from "feudalism" to "capitalism" required the internal changes experienced in England. That is, unless the elite classes can gain full control of production from the laboring classes, increased output will only happen through greater efforts rather than through qualitative breakthroughs.[4]

This chapter examines the social relations of production in Pingxiang County, showing how, like France, a weak state and well-entrenched peasantry led to a stable though sluggish economy that failed to alter the rice and coal markets or transform the county's economy. Powerful lineages extracted surplus production from the peasantry, who harvested two crops per year to survive. In the coalfields, too, miners paid rents for the rights to extract coal, but miners and elites alike still used it primarily for local markets and personal subsistence. When the peasantry employed various acts of resistance to fight off elite exploitation, landlords pushed back through expanded tenancy relations and relied upon state militia to crush acts of rebellion. This, in turn, sparked further resistance from the laboring classes who joined secret societies and rebel bands that spread violence and graft throughout the countryside. As these two strata fought over the social order throughout much of the second half of the Qing era, by the late nineteenth century neither the elites nor nonelites had succeeded in transforming the nature of the relations of production.

"Our ancestors moved to the mountain region"

Migration and settlement in Pingxiang County beginning in the Ming-Qing transition altered the social and political structures. The horrific rebellions that brought down the Ming dynasty in the mid-seventeenth century left much of the interior ravaged and desolate. To try to resettle the most devastated lands and towns, the newly established Qing enacted policies that gave frontier land to people stifled by overpopulation along the Pacific Ocean. The first waves of migrants sought low, flat lands that they could use to harvest rice, as they had in their former homesteads.[5] Once the lowlands were filled to capacity, some migrants moved to the poorer territories and highlands of Central China in search of fields where rice farming could replace less efficient farming strategies. In the years leading up to the fall of the Ming dynasty, the mountainous county of Pingxiang, like the Jiangxi provincial highlands in general, was made up of mostly unoccupied or underpopulated territories and undeveloped peaks and valleys primarily utilized by Daoist priests as well as bandits and smugglers.[6] After the lowlands of Jiangxi Province were taken, migrants subsequently occupied these territories. A mid-Qing-era gazetteer of Yuanzhou Fu, the prefectural area that included Pingxiang County, explained: "Formerly [Yuanzhou Prefecture] abounded in idle land. On account of rapid population increase more land was cultivated but it was still confined to level areas. Since the influx of immigrants from [Fujian] and [Guangdong], their men and womenfolk have systematically cultivated high hills and even steep mountains."[7] Gazetteer entries of incoming families provide a story of a dramatic influx of migration during the Shunzhi and Kangxi reign periods (1644–1723) at the beginning of the Qing dynasty.[8] After this initial invasion, more waves of people continued to arrive, some as independent migrants and likely more as extended lineage members of the first migrants. Population numbers found in the gazetteers—numbers that should never be taken literally but can be used to suggest population change—indicate a population that more than doubled in the nineteenth century from just under two hundred thousand to over a half million individuals.[9]

Those outsiders who began their new lives in Pingxiang County and the surrounding area came from different locations and held varied cultural, linguistic, and social backgrounds. Many arrived with only the items they could carry over hundreds of miles and into the mountains, and they built simple, crude homes and settled as best they could. Given their unfamiliar languages and poor dwellings, local inhabitants often referred to them as either "shack people" (*pengmin*) or "Hakka," a term that literally means "guest people" that came to be used with derision.[10] Shack people often lacked capital and tools when they arrived in the highlands and therefore engaged in slash-and-burn agriculture to survive on the poorer soils not already under cultivation. Many who occupied farmlands and who hoped to recreate their agricultural schemes from the east-coast highlands brought such crops as yams to supplement their rice fields. Increasingly they also farmed commercial crops such as indigo, tea, and *tong* trees, which produced cooking oils. Also, the increased demand for timber led to a great outpouring of migrants into the Hunanese-Jiangxi provincial highlands to supply the cities with highly valued fir and pine trees found throughout the area.[11]

As the economy of the region developed, the social order became more established. Hakka and other migrants who initially occupied the highlands as poor families and communities gradually joined into powerful lineages that dominated much of the county. The most successful members achieved the status of lineage leaders and important landlords.[12] Zhang Guotao, a Communist Party official who was himself a member of a migrant family in Pingxiang County, quoted his mother's description of their Hakka family background:

> In the late Ming or early [Qing], our ancestors moved to the mountain region between the two provinces, although I don't know why. We settled down easily and got along fine even though we were "guest people." . . . Our six families living together included more than one hundred people. We worked hard, farmed hard, and led a frugal life. Not one of us smoked opium or gambled or got into trouble. On the contrary we did much charitable work. Relatives, clansmen, neighbors, and local villagers felt free to visit us often, for we liked and respected each other. Everything went so smoothly that we enjoyed settling there.[13]

This passage provides several important pieces of information regarding the development of migrant families into powerful gentry members. What is most important is that Zhang's mother provides, in a nonconfrontational and self-serving story line, a depiction of the gradual growth of one patriarchy over other relatives and neighbors. From this narrative it is easy to imagine the generations of struggle and negotiation between patrilines over political and financial control of the community. While some succeeded, others failed, and it was the winners who told the story of their means to legitimate power.

In any case, the historical data support Zhang Guotao's depiction of the gradual creation of lineages. The gazetteer informs us that by the Kangxi reign period (1662–1722), many "shack people" in the Jiangxi provincial highlands developed lineages in part to control their landholdings and enhance their chances of survival in the frontiers.[14] Moreover, successful lineages developed lucrative production strategies by buying out poorer lineage fragments and other villagers from different families.[15]

In Pingxiang County, lineages evolved into powerful institutions that controlled land and labor while they maintained order.[16] The local lineage system included both true patrilines as well as fictive relations that were organized by immigrant families to protect themselves and provide access to land and wealth. In an entry in the gazetteer, one author explained that *tu*, or small communities, were organized as lineages and functioned as such:

> The native *tu* of Pingxiang each started from its remote ancestor, and his sons and grandsons continued [to live in the area] for generations. Each *tu* established an ancestral temple and the people of the ten *jia* gathered once a year to worship their ancestors. Moreover, they investigated comings and goings, and prohibited imposters. Therefore, entering the *tu* was very difficult.
>
> There were also guest *tu* to manage the outsiders who for a long time have caused trouble when they moved into the natives' area. Thereupon the clever ones among the guest people assumed [false names] [to move into the natives' areas] to take land and live. Therefore the natives attacked them.[17]

From this passage we can assume that the initial lineage system in the county was fluid and might allow for some to improve their social status by bringing in new skills and being taken into a powerful lineage or through sheer guile and strength. Given Zhang Guotao's description, it seems that these alliances evolved into more solidified groupings by the nineteenth century.

Even as lineages delineated insiders from outsiders, they were also internally hierarchical organizations that separated elites from nonelites. While every member of the lineage possessed a recognized social location, the relative wealth and power of the members differed greatly. Several families used their power as landholders as well as their status within the lineages to become members of the gentry elite and take jobs in the civil service system. Those members who controlled the wealth in the community also became the lineage's leaders. They used the power of land and labor as well as the lineage institutions to manage local affairs. Other lineage members—even those that were given surnames and generation names that mimicked their more illustrious cousins—were part of the commoner and peasant population of the county.[18] This relationship between the elite members and their nonelite counterparts was a contentious one. When goodwill was not sufficient to extract surplus from nonelite members, lineage leaders used brutal force to maintain their leadership and the status quo.

However, beyond naked power, to increase and legitimize their authority local landed elites educated their sons for the civil service exams and thus established themselves as official members of the gentry class. Counties that contained members of the official gentry often succeeded through greater success extracting wealth from the peasantry that was in turn used to hire teachers for educating the sons. Conversely, members of the official gentry used their success and political power to exact greater power over their fellow lineage members. Retired officials were able to invest more into their lineages and the communities as a whole, which further strengthened the polemical and economic power of the community. The majority of the degree-holders came from the southeast coast of the empire, and because of this the most powerful lineages developed there as well.[19] However, even while many counties in northern China and the empire's frontiers held few if any civil service degree-holders, in the nineteenth century Pingxiang County was blessed with a relatively large number of official gentry, both low-level and, impressively enough, upper-level degree-holders as well. The gazetteer shows that in the first half of the Qing dynasty Pingxiang County produced virtually no degree-holders at the highest, or *jinshi*, level and very few at the second, or *juren*, level. However in the Guangxu reign period (1875–1908) the county's inhabitants included fifteen recipients of the *jinshi* and sixty-one holders of the *juren*.[20] The numbers are impressive, indeed. However, it is true that degrees were awarded more freely after the Taiping rebellion to those regions particularly devastated during the years of turmoil.[21] It would seem, therefore, that the county's elite families took advantage of this gift from the court and used it as a means of local power over the peasant population.

Table 2.1 Guangxu-Era Degrees Completed by Family Name

Family Names	*Jinshi*	*Juren*
Li	2	3
Wen	2	8
Xiao	1	4
Liu	0	6

Source: Liu, *Zhaoping zhilue*, 1164–67, 1186–95.

Degree-holders and their agnatic kin engaged in a variety of pursuits based on individual ability and occupational availability. Many official gentry sought out prestigious government offices that provided them with access to the court's coffers as well as personal connections with some of the empire's most powerful officials. Others engaged in various pursuits within Pingxiang County, including nongovernmental or semi-governmental academic positions as well as infrastructural work including building granaries, hospitals, and libraries. Strategies that emphasized activities in the home county indicate a type of isolationist mentality, something almost certainly more prevalent in peripheral counties. At the same time it seems likely that those degree-holders who chose to turn down the chance at public office or those who left office early tended to be those who were less capable than their more outwardly seeking counterparts.[22] Cases like this happened in Pingxiang County even among the most powerful lineages.[23] Even more importantly, as will be discussed later, the important scholar Wen Tingshi left his positions in Beijing and returned to the comforts of Pingxiang County in part to avoid the wrath of the empress dowager and the imperial censorate after he was accused of traitorous acts.

Furthermore, lineage leaders turned to the county magistrates for assistance in their control over the local population and maintenance of the county infrastructure.[24] Located in the county seat's yamen, magistrates throughout the Qing era—many arriving from the more cosmopolitan east coast—helped bring order and civilization to the Jiangxi provincial highland. The gazetteer credited them with building and repairing orphanages, bridges, and libraries, and in addition they also led militia to put down rebellions and assisted in the education of the sons of the county's most powerful families.[25] Nevertheless, magistrate power was limited. Unless they could gain the support of the landlords and lineage leaders, they were unlikely to make progress in repairing and maintaining the countryside. When these two sectors worked in tandem, they were often successful in maintaining the security and economic viability of the community.

Overall, the evidence points to a changing environment from a frontier spotted with clusters of subsistence peasants to a more complex society dominated politically and socially by lineage leaders and their government allies. As the lineage elites gained greater control over the politics of the countryside, they also commoditized land and labor. In so doing the economy gradually moved toward one in which subsistence of the peasantry relied on relationships with the landed elites while the gentry, in turn, relied upon the labor of their tenants for their own survival. At the same time, as will be shown later, while gentry dominance forced commoners to work harder and increase output, elite power was still not sufficient to break apart the patronage system and spark a technological and marketing "take-off."

Landholding Incomes and Tenancy Costs

By the second half of the Qing era, lineage leaders used their wealth to consolidate farm fields and mines under their control, decisions that had both economic and social consequences. Consolidation of property meant greater entrenchment of the power of the landholding families over the peasants who rented paddy fields and mineral deposits for these subsistence labor strategies. Conversely, land-rich lineage leaders relied more heavily on rental incomes for their own subsistence. Under this relationship, both landholder and peasant relied upon one another for their survival.[26]

As we have seen, Zhang Guotao explains that, because of their hard work and frugal living, some members of his lineage became powerful landholders in the years after their migration to the Jiangxi provincial highlands.[27] He also explains that during the late Qing dynasty his ancestors were landlords and gentry:

> For generation upon generation our people had been scholars. We were at the same time landed gentry; and in our vast, rambling house lived my grandfather and his five brothers and the families of all six of them, each with land that supplied an annual income from rent of between five hundred and one thousand *piculs*[28] of grain. Four of these six brothers were qualified to wear Imperial hat buttons and court robes. Two had gained their Imperial degrees by making donations to the State. Abolition of the Imperial examination system began in my father's generation. . . . My father was in the last group of students especially selected to sit for the provincial examinations, which he passed. . . . But only two or three members of the family in his generation took Imperial degrees.[29]

Based on this passage as well as others quoted earlier, Zhang narrates the linkages of kinship ties, landholdings, and civil service degrees in his family's background. He suggests that while his family eventually moved to the city and lived as part of the urban gentry class, his more distant relatives continued to work in the villages and paddy fields.[30] In this way, lineages in the county became more diversified, impersonal, and hierarchical. While each of the members counted on the others for political and economic necessities, the expanding division between the gentry families and their peasant brethren altered these relationships as their status influenced their view of the lineage and the world around them.

This relationship between the gentry and peasant members of the lineage is most tangibly viewed in landlord-tenant agreements and land purchasing contracts that dominated much of late imperial China's agricultural economy. The landholding pattern that dominated Pingxiang County was based on a small faction of landed elites renting land to tenants with lifetime contracts.[31] The percentage of the population owning more than 50 percent of rice paddy fields was 5 to 10 percent, while upwards of 80 to 95 percent of the people rented at least some of their land.[32] Rents were often exceedingly high and amounted to at least 45 to 50 percent of the first crop produced each harvest year.[33] The relationship between landowner and tenant was delineated by a written lease that usually held the peasant to a contract of five to seven years and stipulated the quantity and quality of the rents. The contract often also prescribed any tribute that was to be paid by the peasant for the rights to use the land. For the most part tributes amounted to a single chicken or rooster every year and some predetermined amount of cash for every picul of rice produced.[34] In his report on tenancy in southeastern Jiangxi Province, Mao Zedong explained,

> Eighty percent of the rent in the county is paid in grain; the other 20 percent is paid in money. About half the ancestral, religious, temple, and bridge trusts accept money because most of their tenants are members of ancestral trusts, religious associations, and so on. Because of this peasants can keep the rice for their own use and use money to pay the rent (figured according to the current rice price). Most of these tenants make money through small businesses or by raising chickens and pigs for market.[35]

Mao also suggests that in times of hardship, due to poor farming weather for example, peasants and landlords often agreed to reduce the rate of rental payments on the lands. However, many tenants were coerced into paying fixed rates of grain every year regardless of their harvest size. But during these times, tenants often successfully negotiated with their landlords to accept inferior-quality grain as partial payment of the rents due. This contention is not supported in the proverb cited in chapter 1, which held, "If there is no rain during (the spring) give the land as repayment back to the landlord."[36] This proverb would indicate that the fear of losing all one's possessions after a poor harvest remained with the people of Pingxiang County. While this fear may have been real for many peasants who struggled to feed their families, it would seem that for the most part stories of death sentences passed down by landlords were essentially apocryphal.

In fact even as many people lived as tenants and paid substantial rent, they usually enjoyed lifetime control over the land they worked.[37] For the majority of tenancy relations the landlords and tenants relied upon the two-lords one-field system. Under this agreement, the peasants who tilled the land also controlled the land and had top soil rights. These rights allowed them to farm as they pleased for an indefinite period of time and gave them the authority to sell their rights over the use of the land as they wished. Landowners claimed the rights to rent payments and could sell these rights to other investors if they believed it was in their personal best interests.[38] Under this tenancy system, peasants secured not only de facto ownership of the lands but also fixed rent payment costs. Since landlords no longer needed to maintain a physical presence in the villages, they attempted to add to their incomes through interests on loans, fees, and even some increase on rents due to population pressures and the dearth of available properties.[39]

Just as tenancy in the nineteenth century was part of the patronage system, so too land purchases provided for the continued subsistence of the seller while allowing for the enhanced wealth and substance of the purchaser. As Zhang Guotao explained earlier, migrant families occupied the county's highlands for several generations beginning in the seventeenth century. Over time, landholding families that fell into economic difficulty were sometimes forced to sell their fields to the families with stronger economic standing, a decision that often sent the sellers into even greater economic stress. In some rare cases—primarily when selling gravesite property—the contract stated that the purchase was irrevocable under Chinese law (usually written as *juemai*), giving the buyer clear holding over the property. More often people purchased land as conditional sales (*dian*), meaning that the seller intended to repurchase the land at a later date.[40] Under the terms of the agreements, land that was purchased through conditional agreements was customarily farmed by the seller who in turn paid rents of around 50 percent of the summer crop to the purchaser, thus allowing the seller access to the fields but only providing half of the total output.[41] During the Qianlong era (1736–1796), the court announced that all conditional sales could be revoked by the seller after thirty years if the seller could afford to repurchase the lands.[42] As the county's population grew and the density of the inhabitants increased, the value of the property increased while the immediate needs of some for their subsistence rose. This placed greater stress on marginal families, forcing them to sell. At the same time the poor incomes that landholders gained from their holdings was insufficient to entice them to make large investments in their land when official pursuits promised a greater payoff. Under these conditions, elites tended to purchase land irregularly and in piecemeal rather than single manors.[43]

In Pingxiang County specific data regarding landholding patterns and tenancy relations support the broad outlines described. Land surveys completed by the Communist government in the early 1950s found that a very small percentage of the county's population owned nearly half of all cropland. Specifically, it was determined that poor and middle peasants (as defined by the Communist Party) comprised more than 80 percent of the population and held slightly over 30 percent of the land. Landlords and rich peasants, on the other hand, just over 4 percent of the population, owned nearly 48 percent of the lands under the plow.[44] These numbers certainly indicate that upwards of 40 to 50 percent of the county's farm lands were likely rental properties owned by the land-rich classes and rented out to poor peasants who paid half of their output to the landholding class as rents.

Furthermore, documents of landholdings and purchases within the county reveal that elite holdings dotted the countryside and that rental agreements were probably similar to those throughout the southern inland counties. In particular, lands held collectively by the Wen lineage as described in the lineage records provided the lineage with rental income.[45] Similarly, the land sales I collected from local archives described small parcels of mine property that dotted the mountains around the town of Anyuan.[46] These data support the contention that by the late Qing period, landed elites were gradually collecting many of the county's most lucrative properties and renting them out to peasants and miners. While they were not exclusively interested in landholding for its own sake, it both provided the revenues needed for their gentry lifestyles and gave them greater status and control over the county's people and economy.

For scholars today, determining the incomes of properties and the value of that property is always difficult as the calculations provided in the documents rarely ever match up completely. Moreover, unlike most regions in China that delineate land by the *mu*—a measurement that equals 1/6 acre or 1/15 hectare—the population in Pingxiang County used a local measuring system called the *ba*, or the hand. To make matters even more complicated, the local population calculated the parcels of land differently based on the quality of the land. For the best-quality land, 1 *mu* equaled 30 *ba*; for secondary-quality, 50 *ba* equaled 1 *mu*; for the next level of quality, 80 or 90 *ba* was 1 *mu*; but for the poorest quality land, as much as 100 *ba* equaled 1 *mu*.[47] The county magistrate, Gu Jiaxiang, apparently deemed this system too complicated when he purchased land for the railroad. To establish a standard means to assess the county's property, he sent men out to calculate the *ba* to *mu* ratio. He established a standard that 20 *ba* equaled 1 *mu* of land. Based on these calculations he attempted to determine the output of grain per *mu* of land. For reasons that we cannot know, his calculations are certainly off by a substantial amount. Specifically, Gu's math would indicate an average output of slightly more than 10,000 kilos of grain per hectare, even though most studies calculate that the average southern paddy field produced just over 2,000.[48] On the other hand, the Wen Lineage Genealogy—an undated collection that was probably completed in the 1880s—provides much more likely descriptions of lands held in Pingxiang County. In a depiction of one parcel of land that was 515 *ba* in size, or 1.7 hectares, that parcel brought in an annual rent of 50 piculs or 3,023.85 kilograms of grain. Another property was listed as 500 *ba*, or 1.6 hectares, but took in 40 piculs or 2,418 kilos. The other fields listed for this lineage's temple held similar ratios of land size to rents: 300 *ba* collected 30.1 piculs plus some money, and a small plot of 66 *ba* took in 10.4 piculs. With this property, therefore, the Wens took in at least 130 piculs of rice every year for the maintenance of the temple.[49] Assuming rents represent about half the total output of the main summer crop—a fair assumption for late Qing China—I calculate that the Wen property produced on average 260 piculs or 3,559 kilos of rice per hectare. These fields then produced a second, smaller winter wheat crop or other lesser crop that provided more food for the subsistence of the tenants but the contents of which are not given. These calculations are imprecise to be sure and are slightly higher than the more likely average of about 2,000 kilos per hectare found in the Jiangnan area.[50] In order for the Wen lineage to bring in the levels of income stated in the lineage documents, peasants apparently paid the same type of high-level rents as found in most of southern China.

In sum, migrant farmers in the early Qing era owned fields that they farmed for their subsistence, but population pressures and political intensification fomented a system that forced the peasantry to give up their surplus. The intensification of lineage power in the hands of a landed elite, one whose economic power was buttressed by political legitimacy and imperial connections, subjected peasants to the pressures of tenancy. Not only was this pressure an economic one, but it also meant that peasants relied upon landlords for their subsistence and were beholden at a social and political level for their lives. Even though this peripheral community existed beyond the central markets and political centers of China, the economic and social structures were very much a part of that greater society. Taking this one step further, we can assume that the coercive power of local landholding elites over the subsistence of large swaths of the population followed that found in much of southern China. This level of surplus extraction kept the majority of the population at or near subsistence for generations and eventually sparked the revolutionary conflicts that overturned the society in the early and mid-twentieth centuries. Even more, as peasants were increasingly coerced into agreements with landed elites for access to needed paddy fields, so too the landholding classes had to maintain and expand their relationships with peasant labor in order to ensure sufficient rents to fund their lifestyles. That meant that for the economy to be transformed into a market-driven one, not only would peasant's lives be harmed by the loss of a patronage-based safety net, but landed elites' subsistence could be left vulnerable as well.

Shaking the Money Tree: Mountain Lords and Mining Labor

Having established rental rights over most of the county's croplands, landed elites also began taking direct control over the mines and mineral fields that dotted the county.[51] The highlands of Pingxiang County included land parcels with deposits of minerals and also significant forests of scrub trees and hardwoods utilized for construction of homes as well as mineshafts.[52] By the second half of the Qing era, much of this property also fell under the control of the powerful lineage families who established clientage systems with the local population or recruited working poor from elsewhere to do the labor for pay.

Elite takeover and exploitation of timber, coal, and other minerals developed slowly and unevenly. Whether the leaders became more aware of the large coal markets taking place in the Hunanese provincial river valley or they simply believed that local markets could sustain increased production is not clear. In any case, by the post-Taiping era, lineage leaders viewed coal seams and other minerals in the mountains not simply as subsistence resources held in common but rather as areas of qualitatively new economic development that could be owned and rented for surplus gain. Local gentry viewed the coal deposits in the mines as "treasure bowls" of natural resources that allowed them to "shake the money tree" for large profits.[53] In the previous chapter I explained that during most of the Qing era commoner families integrated peasant labor strategies in coalmines and lumber firms. Even under this premodern economy, mountain output took the form of cash cropping and mineral extraction and therefore relied on the market more than the staples and subsistence cropping performed in the lowlands. Since coal was sold in the marketplace for cash, seasonal miners paid their rents in cash rather than kind like the rice fields described earlier.

By the late Qing era, the value of coal was rising in the marketplace, and this sparked several powerful lineages to engage directly in large-scale coalmining schemes. Just as peasants turned to local lords for paddy fields, many also paid rents to the lords for use of nonfarming property for rights to hunting and timbering in the forests and fishing in the lakes. Similarly, lineage-controlled mineral deposits and mountains required a payment of rent to the lineage leaders who claimed the lands, and over time a clientage system of rent for mineral rights was incorporated into the peasant strategy of subsistence agriculture. Lineages began actively opening mines as early as the Yongzhong reign period (1722–1735). At this time, a gentry member and lineage leader named Deng opened up a mine called Hongzihao (Vast Company) at the place called Jinyushi.[54] By naming the coalmine in this fashion, Deng in essence proclaimed not only that he had founded a mine in what he hoped was a lucrative location but that he had simultaneously formed what can be called a "lineage trust" or a lineage-based corporation.[55] These companies were funded either through individual capital or more likely with the support of lineage ancestral halls and the financial backing of fellow lineage members. Unlike modern firms that employ specialists and seek profits from investments, lineage trusts utilize clientage relationships with laborers and expend the surplus as personal gain.[56] Deng was probably also the first true "mountain lord" (*shan zhu*) in the county, defined as a person who controlled a mountain and became rich from its ownership.[57] The records indicate that another person formed a second smaller mine, but little information is given about its founder or significance. Subsequently, after a lull of some six decades, during the Jiaqing reign period (1795–1819), several lineages established significant mines in the highlands. Local miners Li Shaobai, Gan Chengqing, and Song Zhishou founded four coalmines in undisclosed locations throughout the county.[58]

Finally, over the next several decades other lineages opened mines they hoped would provide them newfound wealth and thereby established themselves as "mountain lords" (*shan zhu*). During the latter half of the nineteenth century, at least fourteen different lineage leaders or trusts emerged as mountain lords. Hoping to extract profits from the local coal deposits, by 1895 they established a reported 265 mines in the Anyuan mining town alone.[59] The gazetteer, for example, describes how Li Youru became a coalmine owner in his later years. After retiring from posts as a Lianghuai salt commissioner and a member of the official bureaucracy in Zhili Province, he returned home to Pingxiang County and sought to use his talents to develop his homeland. His first project engaged in coalmining near the area that was to become the town of Anyuan. In his eulogy, a biographer explained that Li Youru "used Chinese and foreign methods of coal extraction and made profits that were extremely extensive." His desire for profits certainly sparked his first forays into coalmining. However, Li Youru also left the government and became a mountain lord because he hoped to provide more fuel for the community and lineage.[60]

At the same time, while the description Li Youru, a successful Confucian scholar and retired official, portrays the coming of lineage coalmines as a gentlemanly pursuit, in fact it was nothing short of a feeding frenzy. The desperate grab to make mine claims sparked squabbles over property between the lineages leading to feuds in the hillsides. The fighting began in the nineteenth century when Xiao Ruofeng opened another mine at the Jinyushi fields. A gentry member, Xu achieved his *juren* degree in 1837 and went on to earn his *jinshi* degree and take a position in the Hanlin Academy sometime later.[61] He opened the mine after retiring from public office, and when he did, other families attempted to open mines close to his claim so as to receive the benefits of the lucrative site. Xu fought to keep people from other lineages out of his mine area. In the struggle, Xu argued that setting up mines close by and thereby crossing over his seam would be harmful to nature, saying, "If you dig and cut the dragon's veins, you will destroy the *fengshui*." But others established mines in the name of their individual lineages, and those who were not members of that lineage could not mine at the site.[62] Zhang Guotao writes that the mountains around the county "cradled numerous primitive little coal mines, and armed feuds between individual miners and between rival mines were commonplace."[63] Given that struggles over mining properties were also contests between lineages, the violence took on the tenor of family feuds prevalent in southern China in the nineteenth century.[64]

Mountain lords established their claims in the name of their lineages and their output provided for the membership. Some of the most lucrative mines were located in a tight region in the center of the county near the county seat. The Jia lineage owned fields in the village of Anyuan while the very powerful Wen lineage established mines to the northeast in the area known as Zijiachong. A little further to the northeast stood the Wangjiayuan mines of the Zhang and Zhong lineages. To the southeast of Anyuan, the Peng lineage controlled the mines at the location known as Tianzi Mountain. Beyond the central fields near Anyuan, in the easternmost part of the county, the Ouyang lineage mined the fields around Gaokeng.[65] Other lesser lineages controlled still other sites nearby. Those lesser lineages pooled their holdings together into "firms" (*shang hao*) that acted as a single entity.[66] The more powerful lineages used their wealth and status to further their collective incomes. The Wen lineage, for example, also held lands nearer to the western border of the county in Guisheng Township near the town of Xiangdong along the western border. Their lineage genealogy provides a list of mountain sites and rents received that were earmarked for the maintenance of the Northern Ancestral Hall (*Beitang gongci*). Tenants of the Wens were required to pay rent on the land once a year. The data do not describe the size or output of the mountain sites, and thus we are not able to determine the percentage of rent exactions to the total assumed output. In fact, they do not even indicate whether the sites are coalmines, forests, or just poor croplands. Each of the sites is listed as a cliff (*zhang*), an embankment (*tang*), or a slope (*po*). The data do briefly describe each location by name and indicate that the sites received annual rents of anywhere from 1 to 8 cash per year. Whatever renters did on the sites, the list makes it clear that the tenants paid in cash rather than kind as they were required to submit for some of the rented croplands. Based on this list, the total income of the thirteen sites was about 59 cash per annum, a paltry sum given the value of paddy fields and rentals incomes discussed earlier. Therefore, these sites were likely quite small or poorly suited for whatever their use. The data also indicate that the rental properties were scattered among several mountains and named locations, unlike the mines as described earlier.[67] In all, the descriptions of the mountain property indicate that powerful lineages controlled various plots of highland property and that these sites provided commoners with supplemental subsistence properties.

Lastly, it should be noted that not all mines were run by the lineages themselves. Merchants oversaw at least some mining operations, and they, in turn, subcontracted with local lineages for the rights to rent out the property. Under this system, merchants paid workers wages to mine the field and gave a portion of the profits to the landlords as rent. Some of the more successful merchants joined into consortia and pooled their investment capital together. With the added funds, they opened more and larger mines and more efficiently transported coal to the markets. The most successful merchant firms also invested in coking ovens that allowed them to produce higher grades of mineral for sale and sell their product at higher prices than most of their competitors.[68]

In all, the expansion and transformation of mining in the late nineteenth century increasingly forced the workers to either rent mines they had previously accessed for free or to access mines as employees of newly created subcontracting firms run by petty merchants. Lineage leaders and gentry, on the other hand, expanded their hold over the county's economy as mountain lords and gained rents on properties deemed nonessential just a few generations prior.

Owners and managers alike brutalized their workers and provided only rudimentary skills and tools for the jobs. Similar to the contract labor systems found in large farm plantations throughout the empire, mine owners hired contract bosses to recruit semi-skilled and part-time workers for cheap labor power.[69] Under this system the contract labor boss not only recruited but also trained and housed the mine laborers in exchange for payment for all these services by the mine boss.[70] In many cases contract labor bosses created gangs of laborers that moved around the empire in search of employment. The government feared these roaming workers because they were particularly difficult to control. Not only were they likely to act as unruly miners characteristically did elsewhere in the world, but they also failed to live and work within the confines of their natal counties, as desired by the state.[71] Under these conditions, the same landholders who used their economic power to control local commoners could also gain influence over migrant workers interested in laboring at mineral extraction and purification.

Even though China is typically thought to be a case study in labor surplus, mine bosses and owners struggled to recruit adequate labor for their mines. For those mines or firms that utilized labor year-round, securing workers was particularly difficult in the summer due to the high labor demand in the paddy fields.[72] In some cases the need for labor, particularly low-wage unskilled labor, forced the mine owners and contract labor bosses to adopt extreme measures. In one particularly chilling account of mine labor recruitment and control, the magistrate from Leiyang County in Hunan Province, roughly 170 kilometers south of Pingxiang County, wrote of the use of coercion to recruit men into jobs in which they were not likely to survive. I quote the memorial at length because it clearly describes the brutality and intricacies of the mine labor oversight. The passage begins with the magistrate explaining that many of the coal pits required laborers to run water pumps to reduce flooding.

> To manage the water pumps, the mine owners hired foremen, known as "water men." To fill this post they usually picked the worst elements of the local population, men who are extremely violent and wicked. These men, allied with local gangsters, have formed a Blue Dragon Society and accumulated huge amounts of money. To trap poor people, they established gambling dens and sold opium; then they lent them money at usurious rates. Moreover, they colluded with wine shops and restaurants to raise their prices. Badly in debt, the poor people had no choice but to sell themselves to the mine. They would also sometimes capture travelers passing through and force them to work in the mine.
>
> The foremen built near the pit dark, damp earthen cubicles which had only a single opening. Surrounded by stockades, both the entrance and exit of these cubicles were controlled by foremen. These were known as "sealed drums." People lured, bought, tricked, or kidnapped were all incarcerated in such "drums," and were called "water toads." Their clothes and shoes were stripped off and they were forced to work manning the water pumps in alternating shifts day and night without respite. No consideration was given to their hunger and cold. Those who looked tired had their backs whipped, and those who attempted to escape had their feet slashed. Moreover, because it is freezing and the work is extremely heavy, the weaker miners usually died within a fortnight, and the stronger ones suffered from rotten legs and swollen bellies within a couple of months. Without rest and medication, they perished helplessly.[73]

This description not only implies that mine owners could not find adequate laborers but that when they did they mistreated and overworked them. Most small mines in China worked their miners for at least twelve-hour shifts, and many worked even longer than that. Furthermore, the wages paid to miners in the small mines was often meager even by contemporary manufacturing labor standards.[74]

In sum, Pingxiang County's mines were an important site where the divisions grew between those who owned coalfields and hired and brutalized their workers and those who owned no land and therefore took those jobs to maintain their subsistence. Even though peasants owned some land and some of the mines, there was a stratum of the community that established themselves as mountain lords and powerful merchants. Many members of these families achieved the ranks of the gentry elite by passing the exams and entering into the government. These men emerged as among the most powerful in the county and its effective political and economic leaders. And among these county elites, none was more powerful than the Wen lineage and the lineage-based mining firm known as the Guangtaifu Lineage Trust.

The Guangtaifu Lineage Trust—formed sometime prior to the summer of 1896 by two members of the Wen lineage together with the Peng, Zhang, and Zhong lineages—dominated the fields near the village of Anyuan and the mountains immediately to the east.[75] One of the members of the Wen lineage, Wen Tianyi, was a minor lineage leader and mountain lord, and the other, Wen Tingjun, was a significant mountain lord who owned fifty coal mines and was an official-in-waiting. Another source describes an unnamed Wen lineage member who owned the Zijiachong east of the county seat in the spring of 1896. While this reference may be to Tingjun, archival documents from the Pingxiang Coalmine Bureau list several properties owned by another man named Wen Pinshan, and still others list other Wen family members who list the locations of their properties as Zijiachong or the nearby village of Anyuan as well as other locations eventually opened under the modern regime.[76]

Given all the property under the control of the Guangtaifu Lineage Trust, it was certainly the most successful in the county. Wen Tingjun in particular was able to incorporate his own wealth with capital from his lineage members and friends while he hired men from several different lineages to organize mining techniques and strategies, hire boat haulers and coolie laborers, manage labor, and negotiate contracts.[77] By the fall of 1896, the Guangtaifu Lineage Trust was reportedly able to put out some 1,000 tons of coke over a two month period.[78] By the winter of 1896 Tingjun announced that he would be purchasing at least one Western coking oven that could dramatically increase the firm's output to at least 1,000 tons of coke per month. With this upgrade, the firm promised to outpace nearly every coal producer in China except for the fully developed Kaiping Coalmines in Hebei Province in the north.[79]

The flourishing of lineage trusts and merchant firms devoted to mining and other mountain-based production altered the nature of the relationship between landlord and tenant. Increased population pressures and the consequential expansion in local markets led mountain lords to take control over coalmining in the hopes that these investments would expand their use of local labor power and add to their incomes. Lineage leaders hired an increasingly large number of men to leave their fields during the off season to work in the mines. Even as local elites maintained the broad parameters of the clientage systems, they were able to increase total output through capital investment in some Western technology. They did this primarily by intensifying clientage labor inputs rather than altering the social structure of labor, that is, they hired more subsistence-based peasants to work in the off season rather than creating a year-round factory-style production scheme. The lineage leaders and merchants alike chose to increase output almost entirely through added labor inputs rather than modern technological advances.[80] However, as will be made apparent in the next chapter, when the demand for coal grew exponentially in the second half of the 1890s, court officials entered the county and altered the production strategies as patron-client relations were replaced by modern managerial-based labor relations.

Resistance and Negotiation: Using Elder Brothers to Fight Paternalism

Even as the lineage leaders and landlords used their economic power to gain control over local labor, commoners, in turn, struggled against the political regime. Since acts of resistance are often unorganized and appear and disappear with little forewarning, they are often found in the historical documents as entries describing "banditry" or stories of random mass violence when they are covered at all. The Pingxiang County gazetteer documents explosive actions including the Taiping Rebellion and a few other acts of violence as part of the official history of the county. Yet, it is also important to see the smaller, less violent or dramatic acts of resistance as well. In fact, even more pervasive and important for the history of the people of the county may be the ongoing integration of secret societies and the concomitant gambling and opium use that were pervasive in the decades leading up to the opening of the twentieth century.[81]

Given the many large and small rebellious events in Pingxiang County during this time frame, the single most important event was the Taiping Rebellion in the mid-nineteenth century. This empire-wide uprising was one of the first significant uprisings in the county's late imperial period as well as one of its most violent and destructive. Beginning in 1854, Taiping forces entered Pingxiang County, engaged in sporadic violence, then fled government forces by melting into the local population. In 1856, the Taiping general Shi Dakai entered Pingxiang County where his army took control over the market town of Luxi in the westernmost region of the county. His forces and supporters holed up in a gentry family compound until they were eventually pushed out by *tuanlian* militia and government forces. Government-sponsored forces beheaded large numbers of pro-Taiping forces and their supporters in Luxi and elsewhere.[82] These violent events indicated an expanding distance between the commoners who provided support and succor to the rebels and the gentry elite and lineage leaders who recruited a militia reportedly including over a thousand soldiers. Moreover, the violence each side imposed on the other indicates an extreme level of hostility that had been pent up over generations.

Even after government forces quelled this rebellion, conflicts continued to spark between the owners of the mines and those who worked them. These intensified class relations fostered conflicts over working conditions and family needs even as the countryside became increasingly militarized. With both sides licking their wounds, elites strengthened their *tuanlian* militia forces while commoners turned to membership in the local secret societies.[83] In the highlands around Pingxiang County this violence was especially due to the rise of the secret society Gelaohui, or Elder Brother society. After the victories of the Taiping Wars, Zeng Guofan, general and founder of one of the most successful of the Anti-Taiping armies, returned to his native Hunan Province with his armies. While his forces were viewed by many as heroes, they found few openings for success as soldiers or men of action. So, as they returned to their homelands in and around the Hunanese provincial highlands, the Gelaohui recruited them to utilize their military and combat skills to engage in illicit acts and support the society's underground activities, including prostitution, gambling, and salt and opium smuggling.[84]

These wandering men were supplemented by people who tended to be the community's outcasts and destitute, though not exclusively so. Within the Triad society, Fei-ling Davis explained, "a brothel-keeper might keep company with a literatus, a murderer with a policeman."[85] Similarly, society leaders held court at a particular tea shop within the decanal markets on marketing days, and members could go to the shops to meet with the leadership and otherwise engage in secret society activities.[86] By the 1870s, Gelaohui leaders dominated gambling, prostitution, and opium, as well as underground social and political organizations of peasants, workers, and petty merchants in the Hunanese-Jiangxi provincial highlands.[87] Zhang Guotao explains that one method of contact over the countryside was that the Gelaohui was involved in a type of numbers-running gambling scheme. Runners for the secret society went throughout the countryside encouraging people to take chances on their lottery. The practice helped to develop connections between peasants and the lodges, and subsequently, in times of economic difficulty or natural disaster, the throngs of active supporters swelled.[88]

Once the society recruited the local populace into their ranks, the Gelaohui further organized their followers through a series of "lodges." These lodges were divided into two categories. First, the "inner" lodges included an administrative headquarters that allowed for centralized decision-making. The "outer" lodges were regional centers spread throughout the highlands, were used to recruit members, and acted as entertainment and meeting locations for the membership.[89] Organization of the membership rejected the Confucian ideology that they perceived to be based upon prescribed status. Instead, they developed a form of meritocracy that gave all members status based entirely upon their position within the secret society.[90]

The members viewed Gelaohui as a legitimate alternative political institution to the elite leadership of the county.[91] As such, the secret society took on the tasks traditionally thought to be the duties of the gentry elites. Leaders assisted people during times of economic hardship, acted as a banking system for members who wished to have money protected, and provided funeral arrangements for members. Also, the lodges provided several types of judicial functions. For instance, they helped settle disputes between members and assured people of protection from threats of harm. And they held trials in cases in which members were accused of wrongdoing. Unlike those held by county magistrates, these court hearings were viewed by Gelaohui adherents as legitimate because the poorer members of the society were treated fairly and without the complexities of legal jargon used in the interests of the gentry leaders. This judicial system was, however, viewed by the gentry as a usurpation of the government's legal and legitimate authority. In those cases when members were judged to be guilty of wrongdoing, the punishments, as under the county government, were often severe. Certain members of the Gelaohui were recruited to mete out punishment upon their members, a job they performed with great skill.[92]

Beyond the year-round tea shops and gambling places, secret society power and influence was felt in the annual fairs and festivals that celebrated religious occasions. Usually led by merchant leaders of the towns or cities in which they were located, they were primarily established to engage in boosterism for the local community.[93] During the festivals, workers and peasants alike stopped work for the day, put on their best clothes, and enjoyed the community celebration. Workers who were recruited away from their homes returned to their villages to visit family and friends and renew their acquaintances.[94] These celebrations also represented a time when people from the villages and towns, from among the elites and commoners, observed the customs developed in the countryside that many had left. Harvests and other calendrical holidays remained important for those who lived by the seasonal clock and tangentially even for those who no longer did. The New Year's festival held special significance for everyone, as it was the time when chits and debts were called in. While peasants often faced bleak winters after paying their debts, industrial workers similarly found that seasonal layoffs and their bosses' pay schedules sparked their anxiety about the future.[95]

The Gelaohui became very powerful by 1870 in Pingxiang County and the surrounding counties in Hunan Province. According to the Liangjiang governor-general Liu Kunyi, in that year the rebels in or near the Hunanese provincial market city of Xiangtan spread out to other locations such as Zhuzhou, Liling, and Yu counties along Hunan Province's eastern border. The Hunan provincial army and navy fought against the rebels throughout the region and forced the rebels to retreat into other Gelaohui strongholds along the provincial border. The rebels spread into Jiangxi Province in Pingxiang County from Liling County in Hunan. Liu explained that the rebellion did not penetrate far into the Jiangxi provincial territory—territory directly under his jurisdiction—but did affect the border areas.[96] The members who joined together reportedly assembled at the sound of a whistle and attacked Pingxiang City. The urban gentry members built up the walls of the city and enlisted militia to counter the invading forces while the regular army under Liu Kunyi was brought into the county. Battles took place in the markets and around the city. Eventually, the regular forces proved too much for the secret society rebels, and when the army reached the city, the rebels were captured and the rebellion was ended. These victories by Liu's forces, however, did not end secret society activities in the highlands. Further activity continued to disrupt gentry control in Pingxiang County in subsequent years. The gazetteer provides a discussion of a similar Gelaohui-inspired rebellion in 1892 that originated from Liling County and induced violence and unrest in Pingxiang County as well.[97]

For the next several decades the Gelaohui and other bandit and militia organizations continued to engage in acts of resistance and unrest in and around Pingxiang County. For example, Liuyang County—which borders both Liling County in Hunan and Pingxiang County in Jiangxi Province—was a fiercely rebellious center of Gelaohui activity. This county alone experienced at least five rebellions in the last half of the nineteenth century. In most of them, commoners attacked gentry academies signaling the antagonism that was growing between the elites and nonelites in the region.[98]

Conversely, the Hunanese and Jiangxi provincial highlands became more militarized by gentry families increasingly organizing militia to protect themselves against acts of aggression. During times of unrest, lineage leaders—usually with the assistance of the county magistrate—brought together several hundred "braves" who were often called upon to defeat even larger armies of mob violence. These hired killers were likely unmarried males, or "bare sticks," who were in need of an income or were tenants of the gentry families who were called upon to perform militia duties as part of their patron-clientage responsibilities.[99] The years after the Taiping Rebellion clearly continued to be violent ones in the countryside. Friction between elites and nonelites was as strong as or stronger than in the previous centuries, and economic changes were only exacerbating the tensions between those who owned land and held status and those who controlled the means of production.

Conclusion

In chapter 1, I described an economy largely devoid of social classes. Peasants farmed to maintain their subsistence and worked in the coalmines or elsewhere to supplement their family's incomes. To the degree that commoners paid rents to their landlords for the rights to use coalfields, this was viewed as simply another form of tenancy. However, with the development of a more vibrant coal economy and an increasingly voracious gentry class, local elites turned to new forms of income to assure their reproduction. While they could never fully capture peasant labor in the Marxist sense of the word, they were able to intensify the relationship.

Using their own capital and direct managerial control, lineages developed firms that used Western technology or other strategies to increase productivity while they recruited, organized, and managed labor like never before.

Expansion of lineage interest over coalmining occurred simultaneously with increased gentry activity in local affairs and thus over the lives of the commoner population. During the late nineteenth century, elite members gained civil service degrees that set them apart from their commoner neighbors. While the new status allowed many lineages to engage in philanthropy and infrastructure repair, others used their political connections and additional wealth to organize militia armies that were set upon the increasingly unruly commoners. Nonelites who were coerced or enticed to work in lineage trust coalmines experienced this new relationship as yet another reason to oppose the lineage leaders in their county. As they counted on kinship ties for subsistence, they also hoped that government interference would push back against the most exploitative aspects of the relationship. This tension came to a boil in the last years of the dynasty, when both classes hoped for imperial intervention to put a stop to the violence. However, as will be clear in the next chapter, when the imperial government did arrive, the state's plan for the county was one that neither the elites nor the nonelites wanted or understood. At the same time, I show that new avenues of social and economic gain were increasingly possible under the new regime for those willing to walk away from their past lives and community's values and become part of a new order.

3

Self-Strengthening Up Above and Reorganizing Down Below

Industrialization and the invention of mechanized production required the use of managerial overseers who controlled the daily activities of labor and machines in increasingly larger factories. In the United States, independent merchants hired educated men to assist them in expanding their companies and implementing new ideas and polices of management. Socially, this had the effect of creating a new professional class that emerged from growing firms and nascent factories. In China, on the other hand, government officials who hoped to modernize their province's production and distribution sectors hired mangers and experts from within the imperial bureaucracy. Some of these managers achieved their positions through the civil service system and boasted impressive skills in Confucian scholarship, while others were petty merchants who purchased degrees in the hopes of achieving higher social rank. In either case, they were not hired for their skills in new forms of economic thinking or labor organization but were instead called upon to be official bureaucratic overseers of strategic artisanal firms. Under these constraints, the new managers quickly realized that their decision-making powers were limited to those accepted by the imperial state. Even so, for communities such as Pingxiang County, this new bureaucracy was experienced as a new set of tentacles of power that emanated from the court and altered daily lives and economic and social strategies. While most studies of Chinese industrialization have discussed the implementation of this new bureaucracy in vague terms, this chapter provides in detail the give and take and the ups and downs of this strategy of a modern managerial system within the confines of the official Chinese bureaucracy as it affected a specific mining project.

In the previous chapters I showed that Pingxiang County experienced a gradual shift in the systems that controlled coalmining and its financial rewards. In chapter 1, I explained that in the late Ming and early Qing dynasties peasants worked in the nearby coalmines to gain incomes to feed their families while they collected the mineral to heat their homes. In chapter 2, I showed that by the late Qing era the rising population and concomitant rising market forces led landlords and lineage leaders to take greater control of the mining sites and rent them out to seasonal workers just as they rented crop lands. In each case, the bulk of the product was sold locally if not regionally while the consequences of the shift simply intensified the landlord-tenant relationships in the county. However, in the late nineteenth century, as mechanization and modern science altered the production process and expanded distribution, the simple patronage system changed as well. In this way, family-run firms grew into larger corporations too complex for the founders to run on their own, so the owners hired men to act as managers responsible for hierarchical and divisional oversight. These new firms now relied on economies of scale requiring efficient use of natural resources as well as "economies of scope" employing more complex forms of multitasking production that could not be organized through simple subcontracting with intermediaries. Under these conditions, firms became so large and complex that family heads and their extended family members could no longer control all the facets of production and distribution. To resolve this problem, modernizing firms in the United States and some Western countries organized managerial hierarchies to replace family-centered shops. New positions within the firms' hierarchy included managers or their subordinates to organize labor and recourses for efficient production or to make decisions for the firm when the founder was not available or required advice from professionals who were more knowledgeable in organization or the latest technology.[1] In short, modern technologies necessitated reforming the corporate organization and this, in turn, brought about changes in social status and the rights and privileges of employees of the firm.

Even as China implemented technological changes in some of its most strategic industries, the kind of transformative change experienced in the United States simply did not happen in the Pingxiang County Coalmines. Chinese merchants began in the early nineteenth century from a weaker position than those in the United States in that they did not have the political and economic independence American merchants enjoyed. Furthermore, because China's market economy did not employ Western-style capitalism, merchants were undercapitalized and less integrated with the empire as a whole. In fact, even when high demand for certain goods provided opportunities for indigenous companies that competed with modern Western firms, Chinese merchants usually relied on investments in local farming properties that they determined provided more stable rates of return as opposed to untrustworthy investments in small firms.[2]

Because personal investment in manufacturing constituted an unnecessary gamble, Chinese merchants achieved the greatest levels of success as members of the official bureaucracy, such as the salt monopoly or as managers of some of the court's strategic industries. The significance of this is that even as men gained powerful positions and elevated social status within the production process, there were serious impediments to social and economic transformation.[3] In essence the industrial schemes were in reality little more than derivations of government bureaus that dated back two millennia and did not amount to a transformation of the Chinese state or society. Furthermore, because of the lack of corporate initiative and a lack of interests in economic growth, officials created factories and mines in a haphazard fashion that did little to bring about the integrated industrial change necessary for China to emerge as an equal in the global economy.[4]

At the same time, as China's court decided to expand its industrial strength in a set of policies collectively referred to as the "self-strengthening movement," more men found opportunities to enter into the new economic order. Throughout most of the Qing era, low-level gentry purchased degrees that helped secure positions in the official bureaucracy. With the new policy of mechanizing the economy, these purchased degrees provided paths into the empire's strategic industrial firms as well. This meant that ambitious men with useful skills could become professional managers or compradors. Their successes encouraged some commoners and local elites to risk leaving behind the comfort and security of their agriculture properties and petty marketing strategies and enter into the official bureaucracy as managers and compradors. Subsequently, they utilized their merchant and educational expertise to assist the government in creating factories and integrating Western technology with local labor.[5]

In the case of Pingxiang County's coalmines, China's attempt at self-strengthening largely eliminated local production of fuel for personal consumption, replaced with an industrial scheme linked to the larger industrial world. Under the new system, mountain lords and lineage leaders turned away from local marketing strategies as well as their patron-client agreements with fellow lineage members and local tenants.

In the place of subsistence production, coal became a commodity and court officials and even foreign engineers replaced mountain lords as the legitimate owners of the mines and, therefore, mining jobs. When viewed from below, the court's new policy meant that local landlords lost their social and political status as they no longer controlled the mines or the mine jobs. For the community as a whole, the loss of lineage-based jobs in the mines meant a degradation of the patronage safety net and the further atomization of the county. If all these changes do not match the importance and heft of the social changes experienced in the West at this time, they are still important alterations in the bureaucracy and its impact on the entire society.

Late Imperial Reversals on Mining: Hands-Off Becomes Bureaucratized

During the late Ming and much of the Qing dynastic periods, the court imposed two diametrically opposed policies on mining that at first curtailed advancements and then hurriedly attempted to expand extraction. Initially, after a series of scandals in the late Ming era, the court imposed a severe ban on most government involvement in mining altogether. This policy, which continued to a large degree throughout much of the Qing dynasty, may have stifled any "take-off" stage that might have occurred from government investments and managerial oversight.[6] Conversely, once Western technology arrived in China in both violent and nonviolent forms, powerful Chinese officials made an abrupt about-face using their expansive bureaucracies to enter into strategic mineral extraction on grand scales.

The initial hands-off policies were established in the late Ming dynasty after a particularly significant event of political scandal. In the last years of the sixteenth century, court-sanctioned eunuchs used their guile and influence to push aside formal bureaucrats in charge of oversight of mining and taxes. In time, the eunuchs oversaw or even managed mines and ironworks throughout the empire, often over the objection of local gentry. While some succeeded at expanding productivity, in many more cases incompetence, scandal, and the overwhelming tasks at hand led to local opposition to the manipulation by these eunuchs. By 1606, the problems had become so large in Yunnan Province that local military joined with commoner opposition movements numbering some eight thousand people to assassinate the eunuch mining official in the province. Under this violent political pressure, the Wanli emperor began to push aside the eunuchs from their places of power. The court quickly instituted empire-wide bans on government mining and placed restrictions on local firms. These policies stifled large and small mining firms for centuries.[7]

During much of the Qing era, the court continued to place heavy restrictions on mining of most minerals. While some small mining continued either illegally or with tacit support from the local officials, coalmining was not officially approved until the 1740s. After this century-long moratorium, the court specifically ruled that mines had to be owned and operated by the local population, for local consumption. The ruling charged state officials with overseeing local production and marketing and fending off outside investors who might try to monopolize production. Furthermore, to reduce the chance of government corruption, the court stipulated that officials must pay a fixed price, usually higher than market price, for any coal they purchased from local mining firms.[8] These rulings, in essence, took most minerals out of the hands of the state and gave them to local petty merchants who operated mines as their near sole source of income or to large landholding families who held mines along with other fields as part of a diversified portfolio.[9]

Even as the mine policies remained essentially restrictive throughout the seventeenth through most of the nineteenth centuries, individual emperors placed their own imprimatur on the laws making them more or less restrictive. Many court officials understood that arsenals, coinage, and other essential requirements of the empire necessitated mining, and thus a complete ban was simply not possible. At the same time, emperors were often less interested in the economic growth that might come of increased mining, preferring to maintain order by encouraging the stable communities engaged in agriculture rather than the efforts of unruly mining crews and migratory workers. It was in this context that the powerful Kangxi, Yongzheng, and Qianlong emperors each established their own policies that were at times contradictory and based on different principles of virtuous Confucian-based government. In Pingxiang County, for example, even as coalmining constituted part of the subsistence economy, the local gazetteer of the early twentieth century explained that the "Ming dynasty mine tragedy" continued to stifle elite investments in extraction and harmed the mineral market. The author does not provide further direct information regarding this issue, but he suggests that these court policies stifled the county's opportunities at economic success.[10]

Then in the mid-nineteenth century, a series of internal and external military threats abruptly shattered China's position in the world leading to the abandoning of these hands-off policies. Beginning with the first Opium War of 1839, the empire became something of a punching bag for many of the Western powers. Wars and diplomatic losses with Britain, France, Germany, and Russia weakened the military power of the state and ended in loss of territory and bureaucratic control over the populous. The internal rebellions including the Taiping Rebellion, the Nian Rebellion, and the Muslim Uprisings added to China's problems. Under this stress many officials in the imperial government reevaluated their position on government interference with mineral extraction and integration of Western technology. They concluded that the government should reverse their hands-off policies and employ direct control of strategic minerals and manufacturing firms. This reevaluation is usually referred to as the "self-strengthening movement," as it was literally an attempt by Chinese officials to pull themselves out of the doldrums and gain more equal footing with the powerful empires of the West.

To bring about imperial success, these officials argued that simply increasing output of strategic minerals and manufacturing plants was insufficient. They argued that, in fact, China required a full-scale modernization and industrialization project. However, unlike Europeans who turned to their merchants for this type of change, many of China's best thinkers felt that projects of this magnitude could not be completed by the commoners but would require government oversight and power. Classically trained officials, they believed, offered the best hope for organizing the new economy while maintaining the virtues of the state. Moreover, they felt that members of the official bureaucracy could secure the financial backing required for projects too large for local merchant investments.[11]

One of the first leaders to enact a large Chinese-based modernization scheme was Li Hongzhang, the former military commander who fought against the Taiping Rebellion. He became the Liangjiang governor-general and then went on to fill the post of the Zhili governor-general, the most powerful governorship in the empire. Holding this office, he established the China Merchants' Steam Navigation Company in 1872 and the Kaiping Coalmines in 1877 that transported important rice shipments from the southern paddy fields near Shanghai to the northern port city of Tianjin.[12] Li decided that these firms had to be strictly Chinese ventures, and so he developed a strategy that he hoped would literally "seek merchants" as the Chinese-language translation of "China Merchants' Company's" name suggests.[13] Li therefore employed the arrangement known as *guandu shangban*, or official-supervision merchant-management, in which official and merchant activities were kept separate but supported one another toward a common goal. At the top of this system, as the governor-general and therefore the most powerful official, Li memorialized the throne for monopoly rights over a certain economic sector. Once the right was granted, he established a firm and he and his subordinate officials retained the rights of ultimate authority over management decisions as well as the selection of personnel.[14] Furthermore, officials secured taxation reductions and other governmental waivers to hold the costs of production to a minimum. For example, *lijin* fees were reduced for the transportation of copper and cotton to treaty ports by the China Merchants' ships. The state supported these monopolies and tax reductions in order to fend off foreign encroachment on Chinese economic interests. By providing financial privileges to the China Merchants' Company and mining rights to the Kaiping Coalmines, the throne also held off other Chinese firms that could undermine these government-based companies.[15]

While the official bureaucracy held final authority in the broad decisions required of the firm, they, in turn, handed the merchants-cum-officials the rights to make daily decisions and were responsible for the losses to the firm as well as the gains. In each of the industrial projects, the governor-general deputed subordinate officials, compradors, or merchants to act as officials in descending hierarchy: director generals (*duban*), administrators or chief managers (*zongban*), and deputy administrators or associate managers (*huiban*).[16] Moreover, Li Hongzhang incorporated foreign engineers and experts into his enterprises as the technology needed for the modernization schemes was simply beyond the understanding of his Chinese confidants. In all, Li hired Chinese and foreigners alike, gave them orders, and held them responsible for daily decisions and financial dealings with regard to the corporation. Below the two top official-merchant positions, the director generals or administrators often hired a large group of low-level civil service graduates or merchants with purchased degrees into positions as ad hoc managers or deputies (*weiyuan*) to handle the minute actions of the corporation.[17] Thus, Li Hongzhang established an altered government bureaucracy with Chinese merchants and foreign engineers, a system that held on to the basis of Chinese sensibilities of proper governance but buttressed that model with the requirements of the new order.

Overall, the China Merchants' Steam Navigation Company and the Kaiping Coalmines provided successor officials with models of further development in the empire. One of the most important was the brilliant Confucian scholar and Liangguang governor-general Zhang Zhidong.

Zhang originally opposed the kind of modernization attempted by Li until he experienced firsthand the power of the West and saw the need for drastic reforms. While in office in Guangzhou, he acted as the chief government overseer of the Sino-French War from 1884 to 1885. Like the Opium Wars before it, this was another in a string of humiliating defeats for the Qing Empire, one which greatly influenced Zhang Zhidong's thinking about industrial development. He came to understand the serious problems that China faced, and he saw clearly the need for industrialization to both improve the country's economy and its military might. During the war, supplies of weapons and ammunition could not be manufactured by the arsenal located in Guangdong Province. To supplement local materiel production, Zhang had to turn to foreign arsenals and munitions production. His interaction with Westerners and their factories impressed him, and he concluded from this experience that Western influence—if handled correctly—was in the national interest. So he decided to use his influence to help bring about these changes.[18]

Zhang envisioned a modernized world that was squarely in line with his own Chinese-centered Confucian ideology. He argued that when it came to dealings with the West, the Chinese should use the *ti-yong* principle.[19] That is, China should use Western technology, but Chinese culture and ways of life had to be maintained. Zhang argued that the *Book of Changes*, as well as other canonical texts, pointed to the correctness of learning from barbarians under certain circumstances.[20] For Zhang, this approach to modernization was essential to the protection of the empire. However, modernization practices like these and others would be done within the confines of official channels. Zhang's administration consisted of a complex bureaucracy that included men of many specialties and layers of trust and insulation within his personal governing framework. Many of these were men who purchased degrees and were considered officials-in-waiting.[21] As such, they could take on special jobs offered them by superiors in the official bureaucracy. In one study of this managerial strategy, Seungjoo Yoon wrote:

> At the top layer was a coterie of chief document commissioners. They were Zhang's confidants and shared secret information with statutory bureaucrats at the top provincial level. One of Zhang's confidants was commissioned to supervise a specific bureau or an office with the title of supervisor or director general (*zongban*). These chief document commissioners themselves employed their own document commissioners. They were called

> either associate director (*bangban* or *xieban*) or assistant director (*tidiao*). Even though these sub-commissioners normally reported to their immediate patrons, they were also under the ultimate direction of Zhang Zhidong, their chief patron. Below them there also existed other document commissioners or translators of various functional levels.[22]

Employing a personal bureaucracy that may well have reached upwards of one thousand expectant officials, Zhang petitioned the court for permission to establish an ironworks in the provincial capital city of Guangzhou in order to supply the arsenals and naval shipyards.[23] To this end he spent more than 5.6 million taels on equipment made in England purchased with a combination of government funds and coerced merchant investment capital.[24] The ironworks was not only to be used for shipping and military needs but was envisioned by Zhang to alleviate transportation concerns as well. While several Western corporations had established railroad lines along the Pacific Coast and within their colonial territories in China, the Chinese empire could not receive the benefits of these enterprises placed on their own soil. To this end, Zhang Zhidong further memorialized the throne requesting the right to construct a railroad trunk line that would connect the city of Beijing with the southern cities. Zhang suggested that the line would pass through the interior of the country and therefore not interfere or compete with the lines already built along the Pacific Coast by the Western powers.[25] He argued that the railroad would strengthen the Qing military, but more importantly, he explained, it would expedite the conditions for economic development in the empire's interior. Specifically, the steel used for the rails could be manufactured from Chinese ironworks that used European technology.[26] He conceded that construction of the railway would destroy the livelihood of some thirty thousand transport workers and negatively affect at least as many more. However, he assumed that the railroad would hire some of those transportation workers back and that the savings from more efficient and less corrupt transportation tax collection would greatly improve the economy in the heartland of the empire.[27]

Zhang Zhidong proposed that the first leg of this railroad line should link Beijing with the Wuhan Cities in Hubei Province, while construction on the southern half of the rail line would begin at a later date.[28] The throne agreed to this proposal and decided that Zhang Zhidong himself should oversee the railroad's construction and that construction should be done from the Wuhan Cities in Hubei Province. Therefore, he was abruptly removed from his post in Guangzhou and deputed to be the Huguang governor-general in Wuhan, a post he held beginning in August 1889.[29] In the meantime, the steel-producing equipment arrived in Guangzhou about one year after Zhang ordered it from the English manufacturers.[30] When the new Liangguang governor-general Li Hanzhang, the older brother of Li Hongzhang, refused to take on the Herculean task of founding the steel mill Zhang had planned, the governors-general, together with the court, decided to ship the steel-making equipment to Wuhan to be tended to by Zhang Zhidong in his new location.[31] And so in 1891 Zhang established not the Guangzhou Ironworks but the Hanyang Ironworks. This plant remained the first and only Chinese-owned iron and steel works in all of China for several decades.[32] And as its founder, Zhang ran the entire operation using only the men deputed within his administration. The leadership of his newly formed ironworks bureau consisted primarily of sitting officials, while their immediate subordinates were expectant officials with backgrounds in Western language and Western technology and industrialization.[33] To supplement his Chinese men, Zhang hired Western experts, including German and English engineers, to help modernize the ironworks.

Even as Zhang Zhidong had experience running the bureaucracy that he intended to employ, he did not understand the complexities of industrialization. And this lack of knowledge quickly led to a series of mistakes, problems, and failures. Zhang's first heedless decision was that he chose to build the ironworks in the small, walled prefectural capital of Hanyang, a market town that was part of the Wuhan triple cities. The site he chose was too small and had terrible drainage problems and poor port facilities for mineral transportation. Furthermore, he constructed the plant before he knew where the minerals were going to come from. Thus, he found out too late that the plant was located too far from iron ore and coal deposits to be economically viable. However, he made this decision in part because he did not understand these technological issues and in part because he feared that the corruption and inefficiency of government officials outside his immediate oversight might scuttle the project.[34]

Along with placing the ironworks in a poor location, Zhang's second mistake was that he purchased the machinery before he secured adequate resources. Initial attempts to construct and maintain the furnaces led to problems Zhang could not overcome, while the lack of adequate fuel supplies and other problems forced shutdowns and added annoyances. After many fits and starts, by 1893 the initial construction of the plant was completed. The Hanyang Ironworks included two English blast furnaces and a Bessemer steel furnace. However, even after he received the machinery and successfully fired up the furnaces, Zhang had not yet secured dependable sources of coal, iron ore, and other natural resources he needed. To begin to solve this problem, he sent foreign engineers throughout Hubei and adjoining provinces to search for suitable mineral deposits.[35] Fortunately, he located an adequate supply of iron ore in Hubei Province at the Daye Iron Ore Mines. These deposits were owned by the official-entrepreneur Sheng Xuanhuai who was in charge of the China Merchants' Company, the firm founded and governed by Li Hongzhang. Zhang and Sheng came to an arrangement for the iron ore mines, and thus Zhang purchased the fields in the same year.[36]

However, obtaining coal and coke needed to fuel the plant proved to be an even more difficult problem, one that Zhang did not entirely solve on his own. As early as 1890 Sheng Xuanhuai told Zhang Zhidong that the Daye iron ore fields included some low-grade coalmines that could be used for the time being.[37] Two years later, Zhang wrote of these fields describing them as assets he could use for his purposes. In the same letter he went on to explain that the coalmines in Ma'anshan, a field located in southern Hubei Province, contained large quantities of coal that was already being mined and shipped to the Hanyang Ironworks. He explained that he planned to enlarge the mines by the summer of 1893 to fulfill the needs of the ironworks, and other letters written about the same time agreed with this assessment.[38] Indeed, by that summer, an unnamed Hanyang Ironworks administrator (*zongban*) explained in another letter that the Ma'anshan Coalmines were beginning to show some success and that the purchase of electric Western ovens was being considered to coke the coal for the ironworks' furnaces. The author also stated that further investment would prove beneficial.[39] However, once they fired up the furnace in the Hanyang Ironworks using Daye County coal, it became obvious that the coal from these fields contained far more phosphorous and sulfur than that found in England and could not provide the sustained heat required using English techniques.[40] Consequently, because neither the Daye nor Ma'anshan coalfields were sufficient, Zhang attempted to supplement his local mineral holdings with fuel from elsewhere. Along with some purchases of coal from as far away as Manchuria and even England, by the mid-1890s he purchased coal from the Western-run Kaiping Coalmines located in northern China.[41]

At the same time, Zhang continued to search for local sources of coal, but he determined that most were unsuitable. Among the many mines tested, by the early 1890s Zhang learned of one potentially suitable coalfield in Pingxiang County, Jiangxi Province. He probably heard of these mineral deposits from a man named Ouyang Bingrong, a resident of Pingxiang County. Ouyang was a promising scholar, and his wealth and status, which probably included kinship ties with the Ouyang lineage mountain lords of the local Gaokeng Coalmines, made it possible to secure a purchased degree.[42] In 1890, Ouyang worked for Zhang as an inspector in the Hanyang Ironworks. In 1891, Zhang ordered him to begin inspections of coal deposits in Pingxiang County, Liling County, and the neighboring regions.[43] The next year, he wrote to Zhang of the coal resources in his county and offered his services as a facilitator in providing coal for the ironworks. Zhang agreed to this plan and commissioned him to establish a government-sponsored coal bureau in Pingxiang County. Ouyang immediately began to purchase coal from local miners, including the Wen lineage, and arranged shipments of the coal to be delivered to the city of Hanyang.[44] By the summer of 1896, Ouyang shipped about 1,000 tons of coke to the Ma'anshan Coalmines to be mixed with locally produced coke to fuel the Hanyang Ironworks' furnaces. Even though these shipments could not meet the requirements of the ironworks, German engineers determined that the coal was superior in quality to the Ma'anshan coal deposits. Subsequently the engineers mixed it at a rate of 70 percent of Pingxiang coal to 30 percent of coal from local sources.[45] While the engineer viewed this find as a possible solution to their problems, Zhang gave up his hopes that yet one more coalmine might hold the solution to his problems.

In fact, Zhang's countless dilemmas with the project defeated him personally and left him overwhelmed by debt. As early as 1892, he wrote to Li Hongzhang that he was using 2 million taels from the Board of Revenue, as well as 800,000 taels from the Hubei provincial government, to support his ironworks.[46] He also took more than 1.8 million taels from his arsenal and textile factories and close to a million more from the imperial salt tax and the defense coffers.[47] Finally, after these attempts to maintain the plants, in the summer of 1895, Zhang again asked the board of revenue for an additional infusion of funds. This time, however, the board refused, leaving Zhang with no options but to turn to a new source of financial backing.

With almost no new sources of capital, poor-quality iron ore, and uncertainty with coal supplies, Zhang was forced to turn to investors and powerful merchants who could assist him with his financial problems, his managerial concerns, and his problematic fuel issues. Zhang's Confucian teachings had left him distrusting of merchants. However, coming from a position of weakness, in May 1896 he reluctantly turned to one of the most powerful and shrewd official-merchants in all of Qing China, Sheng Xuanhuai.

Sheng Xuanhuai's Managers

Sheng Xuanhuai was an official and a member of the gentry whose merchant and managerial experience as well as his personal wealth and physical holdings made him uniquely qualified among Chinese to assist Zhang Zhidong in the mid-1890s. Prior to working with Zhang Zhidong, he acted as a manager and director over Li Hongzhang's industrial empire. As Li's assistant, Sheng developed skills, connections, and properties that were important to Zhang and led to the two men working together.[48] The first reason that Zhang turned to Sheng was that, broadly speaking, working with Li gave Sheng the know-how required to oversee and integrate his industrial scheme. In particular, Sheng developed contacts with potential entrepreneurs who helped him finance certain aspects of the project that came to be known as the Hanyeping Coal and Iron Company. Second, Sheng accumulated assets, including the Daye Iron Ore Mines, that served as a centerpiece in Zhang Zhidong's industrial empire. He acquired these mines with the assistance of foreign engineers in 1877, two years after he was deputed by Li Hongzhang, then–Zhili governor-general, to attend to the development of iron and coal extraction in Hubei Province. Third, Sheng and his merchant and manager subordinates held close ties to foreign investors and experts who could not only provide engineering advice but also desperately needed capital to improve the plants and mines that were the linchpin of Zhang's industrial scheme.

Far from a disinterested party, Sheng Xuanhuai closely studied Zhang Zhidong's industrial empire in the early 1890s, even as he was still employed by Li Hongzhang. As early as the first months of 1892 he began receiving detailed updates of Zhang's progress from a man named Zhong Tianwei, one of Zhang Zhidong's secretaries who also worked for Sheng.[49] Much of the information Zhong passed to Sheng, and several of the documents he attached with his own letters contained detailed analysis of the mines and factories from some of Zhang Zhidong's most trusted advisors.[50] The letters indicate that Sheng knew of the various strategies Zhang employed, including use of the Ma'anshan Coalmines and purchases from the Kaiping Coalmines in northern China. In particular, after briefly discussing these coalmines with Zhang Zhidong in a letter dated December of 1892, he received a letter dated June 13 the following year that included an attached memo from the ironwork's director general (*zongban*) describing coal extraction, purification, and transportation at the Ma'anshan Coalmines. Sheng, in turn, utilized this information in his correspondence with Zhang regarding financial sources and strategies of future industrialization schemes.[51] Specifically, after agreeing to sell Zhang iron ore from his mines in Hubei Province, in 1892, he offered to locate investors who could provide capital for the firm. Because Zhang opposed merchant activities in his industrialization scheme on principle, he initially declined the offer.[52]

Yet in 1896 Zhang Zhidong realized he needed outside help, and so he reluctantly asked Sheng if he and his contacts would save his firm. Even though the letters from Zhong Tianwei prove Sheng was deeply interested in the project from the beginning, he feigned ambivalence. He realized that Zhang's problems provided an opportunity to secure wealth and power from the desperate governor-general and used that leverage to his greatest advantage. Sheng offered to take control of the Hanyang Ironworks but only if he was also made the Hanyang Ironworks director general and director general of the newly formed imperial railway administration. These positions promised to secure Sheng complete control over the court's railroad construction projects as well as directorship over its most important source of iron rails production. Zhang very reluctantly agreed to these arrangements and placed Sheng in charge of his entire industrial firm.[53]

Upon taking the reins, Sheng quickly got to work expanding mineral extraction and transportation in the hopes of increasing production in the Hanyang Ironworks. Even though the Daye Iron Ore Mines were operating under Zhang Zhidong, Sheng developed the iron ore mines further. He helped to construct a railroad line that greatly facilitated transportation of the mineral from the mines to the Yangzi River port, which in turn allowed his workers to load the firm's river barges with extracted minerals and deliver the supplies to the city of Hanyang. Moreover, Sheng was instrumental in solving Zhang Zhidong's problem with accumulation of coal needed to run his plant.[54] In May 1896, while trying to line up investors, he wrote a prospectus listing the assets of the firm. In this letter Sheng does not mention the Pingxiang coalmines but does present data on both the Ma'anshan and Kaiping coalmines. In a subsequent letter dated June 26, Sheng indicates probably for the first time his knowledge of the use of Pingxiang County coal. In the letter, he tells the German Hanyang Ironworks superintendent whose Chinese name was Depei that coal purchased from the Kaiping Coalmines could be cheaper than coal from Pingxiang County because the latter's transportation costs were excessive.[55] Two days later, an official named Yun Jixun wrote a memorial to Sheng from Pingxiang County indicating that Zhang Zhidong had sent him and others looking for the best coal deposits possible. To this end, he explained that he had inspected the coalmines in Pingxiang County and found that they contained the lowest phosphorous content and were purer than the Ma'anshan Coalmines. After bemoaning the drudgery experienced by local miners, Yun speculated that the employment of Western technology in Pingxiang County's coalfields would dramatically improve the output of the mines and therefore provide the ironworks with many tons of fine-quality coal.[56]

While the initial correspondence regarding the ironworks and the Pingxiang County mines indicate that the leaders of the industrial scheme in the early 1890s were primarily officials and degree-holders hired by Zhang Zhidong, perhaps Sheng's greatest gift to Zhang's industrial empire was his collection of experts, engineers, and foreign advisors he put together to begin to transform the ironworks from a mere collection of buildings and mines into a modern and integrated industrial firm. Employing these men of varying abilities and backgrounds, Sheng put together an administration superior to Zhang Zhidong's when it came to organizing and managing a modernization scheme of this magnitude. While Sheng at times drew criticism for relying heavily on corrupt cronies and men referred to in the Shanghai magazine *Shibao* as "protégés and the relatives of his concubines (whose) only forte is enriching themselves," at the same time, the history of the Pingxiang County Coalmines tells us that modernization and industrialization experienced in this isolated highland community occurred due to a concerted effort on the part of Sheng and his associates.[57] Mimicking the generalist-turned-specialist managers and overseers found in nineteenth-century America, these men incorporated Confucian-style professional governance with modern scientific knowledge to recreate and reorganize Pingxiang County's landscape for the purposes of modern extraction and distribution of minerals for the Western plants in the city of Hanyang.

Once Sheng's men arrived in Pingxiang County, they began investigations into local excavation, paying particular attention to the efforts of the Guangtaifu Lineage Trust, by far the largest and most productive coalmining firm. Based on the early advice of Zhang Zhidong's advisors, Sheng initially decided to continue lineage-based mining in Pingxiang County in hopes of supplementing mineral extraction with coalmining done elsewhere. In May 1896, Sheng Xuanhuai asked Wen Tingjun—a member of the powerful Wen lineage, a Pingxiang County mountain lord, and leader of the multilineage Guangtaifu trust—to assist in the expansion and modernization of mining in his home county and to work out plans to ship thousands of tons of the county's minerals to the ironworks in Hanyang.[58] Wen, an expectant official in Jiangsu Province, was accompanied to Pingxiang County in late June 1896 by Xu Yinhui, another Sheng appointee similarly awaiting a position in Jiangsu Province as an assistant magistrate. Sheng apparently hired Xu to both assist the Wens and to watch over them. Upon arriving in Pingxiang County, the two men gained further assistance from fellow Wen lineage member, Wen Tianyi. These three men were called upon by Sheng to expand the Guangtaifu Lineage Trust and to purchase as much coal from other mines and mountain lords as possible to be sent to the Hanyang Ironworks. Almost immediately, however, the other lineages in the Guangtaifu consortium chose to turn down Sheng Xuanhuai's offer. They sold off their shares of the trust and continued to mine for local sales only. Seeing the Wen lineage leaders as the only large mountain lords remaining in the firm, Sheng's men constructed a lineage temple for the Wens located just outside the county seat and declared this to be the new official headquarters of their newly formed Mine Bureau. From this point on, the Guangtaifu Lineage Trust and the Wen lineage became essentially one and the same entity.[59]

By October of 1896, Wen Tingjun and his fellow lineage members began to put into place more efficient European electronic coking ovens, capable of purifying coal much faster than the controlled wood-fired coking ovens used by the local miners.[60] In this manner, they could coke much of the coal before it was sent down the river, a procedure that reduced the transportation costs and promised to assure the ironworks of a higher-quality coal. They placed the first oven at the base of the Zijia Mountain, the mountain that had been primarily owned by the Peng lineage, one of the primary investors in the Guangtaifu trust, and a second was all but finished by the end of the month. The Wens then began to set up similar ovens in at least six other mountain sites, including Gaokeng and Anyuan mountains. They asserted that by the late fall of 1896 the coking ovens would be operational, and they would be able to put out at least 1,000 to 2,000 tons of coke per month, an amount that neared the demand of the Hanyang Ironworks at this time.[61]

As the Wens began implementing the new technology utilizing lineage labor and management, Wen Tingjun slowly gained Xu Yinhui's confidence. In his correspondence to his superiors, Xu never mentioned his feelings about Wen, though he did on several occasions express frustration over the incompetence and corruption of individual officials and merchants in Guangtaifu and among other individuals in the region. In one of his early letters, Xu wrote that he would certainly watch over Wen Tingjun's appointees, who included friends and family members, and promised to fire any whom he felt were undermining the enterprise.[62] Xu must also have supported most of Wen's assertions that the problems of early production were largely beyond his control. In one letter dated September 6, 1896, Xu and Wen told the Hanyang Ironworks' administrator Zheng Guanying that coal production had been hampered by unusually heavy rains that destroyed the kilns, problems with shipping, corrupt merchants, and sabotage by local residents opposed to the coal project.[63] They explained to Zheng that they planned to develop more mines and argued that they could solve these initial problems.

Even as Wen Tingjun expanded and improved the Guangtaifu lineage trust, his lineage brother and top-level *jinshi* civil service degree-holder Wen Tingshi arrived in the county from Beijing in order to avoid court intrigue. In the capital he had been a member of the Hanlin Academy, the court-run center for China's finest scholars, and developed close ties to the emperor and several court leaders. However, his frustration with the war against Japan and his perception that the court was incompetent if not corrupt led him to join the Qiangxue Hui, a political study society that called for greater modernization of China's economy and political structures. This organization included among its members the powerful liberal reformers Liang Qichao, Kang Youwei, and Tan Sitong, as well, reportedly, as Zhang Zhidong himself. Political infighting in the capital over the war put Wen Tingshi along with some of his fellow members in serious political trouble. In retribution for Wen's criticisms of the empire's military policies, the court subsequently stripped him of his position in the academy. Fearing for his life, Wen Tingshi fled Beijing and headed toward home where he hoped to seek refuge. Along the way he stayed with a local official in Shanghai where several documents he wrote that were critical of the empress dowager's policies were discovered by his political enemies. Wen Tingshi was denounced by a local censor in the spring of 1896, and he lost his official position. In political retreat, Wen quickly returned to his home county.[64]

When Wen Tingshi arrived in his home county, some of the local lineage leaders asked him to facilitate various business arrangements with Sheng Xuanhuai's men.[65] Since the local gentry knew Wen to be a man of particularly high standing and ability, they hoped that he could manage the mine in their interests. And for a brief time in 1896 it appears that he did take care of many of the details. In particular, Wen Tingshi acted as the chief delegate for the Guangtaifu Lineage Trust. He met with Sheng Xuanhuai's managers, he was involved in several important planning meetings, and he sent several memos to the Hanyang Ironworks officials. In all, Wen Tingshi argued to Sheng's men that he and the other leaders of the Guangtaifu Lineage Trust were overseeing the coalmining in the county and attempting to solve problems as they arose. They were obviously being pressured from Hanyang to deliver more coal than they had ever previously produced. However, initial shipments were late and of poor quality when they arrived at all. In the summer of 1896, when the first coal was extracted in Pingxiang County for the Wuhan steel mills, transportation into and out of Pingxiang County was limited to two methods, overland using coolie labor to the Hunanese provincial market city of Xiangtan or by boat on the Lu River to the Xiang River, the major waterway that emptied into the Yangzi River at the Wuhan Cities.[66]

Riverboat transportation was cheaper, and the leaders of the Guangtaifu Lineage Trust knew of problems with the boaters and tried to solve them. In a letter based on his initial investigations, Xu Yinhui noted that many of the boat haulers were not from among the county's more seasoned boat haulers who moved tea and other goods to the markets along the Xiang River but were instead little more than impoverished commoners who had recently gained employment with Guangtaifu to stave off starvation. In fact, Wen Tingshi conceded that some of the first boat haulers they hired were not properly performing their duties and that some of them added heavy shale to their loads to increase the weight and thereby raise their pay. To address this issue Wen Tingshi personally sent guards to watch the boat haulers and investigated the local kilns to determine the quality of the coal being produced.[67] Similarly Xu Yinhui found that some of Guangtaifu's boat haulers used their employment with the lineage trust to evade taxes. Because these boats hauled minerals for the imperial-sanctioned Hanyang Ironworks, haulers received a special dispensation from local transport taxes imposed by the imperial *lijin* bureau. However, Xu found that in some cases the boat hauler sought tax exemptions to which they were not legally entitled. He noted that government regulations forbid boat haulers from taking passengers but that some brought travelers through the *lijin* checkpoints while claiming their entire boat and its contents were tax exempt.[68] Xu realized that they were engaging in attempts to defraud the government, and he asked Wen Tingjun to turn to other boat haulers for future transportation arrangements.[69]

Conversely, while boat haulers attempted to defraud the *lijin* bureau, the most egregious matter hindering transportation had to do with *lijin* bureau officials in Hunan Province who used their positions to attempt to gain added "squeeze" from the lineage trust. When boat haulers from the Guangtaifu Lineage Trust arrived at the Xiangtan *lijin* bureau, these corrupt officials falsely informed the boatmen that they had not received notices of the special tax exemption for the coal shipments and that the paperwork on their bills of lading was incomplete, suggesting that they needed to pay the full tax before they would be allowed to pass downriver. Xu was furious when he found that a corrupt official named Li Anjie held upwards of one hundred boats by claiming that the boat haulers had to pay the full tax. The Xiangtan County magistrate, upon discovering the corruption, sent out a warrant for Li Anjie's arrest and placed a notice on the front of the *lijin* office building calling for a major overhaul of the office and naming Li as among the officials guilty of corruption.[70] In his letters to the Hanyang leaders Xu complained bitterly that the bureau had to be reformed if the ironworks in Hanyang was to receive the coal it needed.

Xu Yinhui explained that transportation was also hindered by the local climate and rice farming practices; irrigation along the Lu River meant that year-round boat hauling posed a series of problems. He found that more than 120 dams on the Lu River impeded the progress of the boat haulers. During the dry seasons in the autumn and winter, dams maintained the water-level and controlled the current. When the dams closed, boat traffic essentially came to a standstill.[71] These dams, Xu argued, were essential to the livelihood of the local inhabitants and therefore could not be removed. While he suggested that boat traffic could be improved by dredging up the river bottom to deepen the channel and reduce turbulence, he believed that this would harm local farming practices, and this led him to advocate finding an alternative method of transportation. To this end, Xu mentioned that the construction of a railroad would be an ideal method of transporting the mineral out of the Jiangxi provincial highlands but acknowledged the initial costs would be exorbitant.

In sum, by the end of 1896 Xu Yinhui and the other officials sent to Pingxiang County determined that the Guangtaifu Lineage Trust might be able to satisfy the needs of the Hanyang Ironworks but was hampered by premodern technology and political and managerial corruption that caused bottlenecks the firm could not overcome. Sheng Xuanhuai realized that by employing Wen Tingjun and Tingshi to oversee coalmining and shipments to the Hanyang Ironworks, he could bring in the coal he needed without altering the political and social structures that dominated the county. Since both men held official degrees, they also held the official status upon which Zhang Zhidong and his predecessors relied. However, employing the Wen lineage as the main source of Pingxiang County coal for the Hanyang Ironworks presented serious problems. Almost certainly the most significant drawback was that the Wens could not expand their lineage-based operation far enough to fulfill the quantities required at Hanyang. Even though these men were among the wealthiest in the county, the capital they could invest in the operation fell well short of that needed for modern industry. Furthermore, the lineage-based system that had functioned throughout China for centuries as the common method of organization for large firms simply did not allow for significant managerial breakthroughs required for such a large scheme.

Moreover, even as the Wen lineage leaders viewed the sales of thousands of tons of their county's coal from their lineage's mines as an opportunity to make profits while assisting the empire, several other powerful lineages in the Guangtaifu Lineage Trust apparently opposed the plan to sell their coal to the Hanyang Ironworks and instead continued to sell their coal on the local market. Perhaps out of loyalty to their county, or fear of the coming of foreign influence, these other lineage leaders continued to stymie the wishes of the empire. Since the lineages had held the coal for several generations, if not longer, and the peasants had mined it and fired it in their homes for centuries, the sudden demand by the imperial government for their mineral had the potential to dramatically alter the local politics of the county. While the local population might find this new demand a potentially lucrative opportunity for their community that could be exploited and fostered, Sheng Xuanhuai and his subordinates simply viewed the people of Pingxiang County to be a hindrance to their plans.

To get around the barriers created by local leaders, Sheng sent another assistant to investigate alternatives to the Wen lineage mines. This second man was one Lu Hongchang, a man who apparently had engineering and geological training. Sheng wrote a letter to Lu on July 24, 1896, ordering him to investigate the Xiang River valley in Hunan Province in search of suitable-quality coal at a fair purchase price. After traveling throughout the markets and mines of Hunan Province, Lu was scheduled to arrive in Pingxiang County. The letters suggest that even before arriving there, however, he met up with Wen Tingshi, who tried to convince Lu that the Guangtaifu Lineage Trust was doing good work and could accommodate Sheng Xuanhuai's needs. Wen and Lu briefly discussed an idea in which the firm could be officially maintained as a lineage-led trust but secretly controlled by Sheng's officials. In a letter to this effect, Wen seemed to be already grasping for some type of agreement by which Lu might allow the Wens to keep their firm.[72] And indeed, upon his arrival in late December and after some initial investigations, Lu Hongchang wrote a letter that reached the Hanyang Ironworks by the last days of that month, explaining that he had visited several mines in the central part of the county near the town of Anyuan where the Guangtaifu Lineage Trust operated and that he was impressed by their initial efforts.

He then traveled to the northern coalfields of the county, the region around Anle Township including the market called Shangli. There, he was taken around by Ouyang Bingrong, Zhang Zhidong's appointed official who founded and administered Zhang's Mine Bureau in the county. In the northern part of the county, Ouyang worked with one Yang Shouquan, a native of the neighboring Hunan Province and a former mine owner from the Anle Township area. The special relationship that each of the men had with Anle Township led them to purchase a large portion of its coal.[73] Because of Ouyang Bingrong's position in Zhang Zhidong's administration, court policies required him to purchase coal at the government-mandated—and artificially inflated—prices that threatened to cripple the ironworks. If the plan was to work, the firm was going to need a new arrangement with the coalminers in Pingxiang County.

To this end, Lu Hongchang attempted to assist Ouyang in finding an alternative to the mandated purchasing agreements. Ouyang, whom Lu referred to as Shangli's manager (*bangdong*), convinced him that the Guangtaifu Lineage Trust successfully pooled their capital together in order to combine many of the coalmines in the vicinity to improve output. However, Ouyang and Yang argued that the purchase of the coalmines in Anle Township in the north and some mines in the county's central mountains—mines not owned by the Wen lineage or other powerful lineages in the county—would be better suited to an expanded mining scheme. In a series of letters offering his assessment of the local conditions written in the spring of 1897, Lu contended that even though the Guangtaifu Lineage Trust had produced some coal of sufficient quality, it relied on many mines in poor condition and not capable of producing the required quantity and consistent quality. He also complained that the Wens were paying a very high fixed price for their coal shipments, hindering Hanyang's fiscal concerns.[74]

To address these problems, Lu suggested that he could implement alternative plans to collectively marginalize the Guangtaifu Lineage Trust, plans, as will be apparent, that succeeded in their aim. First, he began the reorganization of some of the county's mines not controlled by Guangtaifu as suggested by Ouyang Bingrong and Yang Shouquan. He purchased and began operating mines in Anle Township, the northern region of the county that included the town of Shangli. This plan failed for several reasons, perhaps primarily because transportation from some of the mines to the ironworks proved to be difficult. Either coal from the northern mines needed to be transported over land to the Guangtaifu ovens before being shipped down the Lu River to the Hunanese river basin or it would have to be sent down the smaller Shang River to Liuyang County and from there to the Xiang River below. He felt that both of these routes were costly given the state of the infrastructure at the time.[75] Also, Lu found that the smaller coalmines in Anle Township contained a higher sulfur content. If coal from the northwestern sections of the county were to be used in the future, he would need to expand and improve the mines and address the local coking process. If he could not succeed at these changes, he surmised, mining there would be more expensive than the Guangtaifu trust mines. Lu did mention that he found one mine in the area that contained high-quality coal and that for about 400 to 1,000 cash they could extract and coke coal from that site. He decided to buy one of the three shares of this coalmine and planned to buy at least one more share if repairs proved successful.[76]

Lu Hongchang also began to buy up quality coalmines in the area where the Wens had established their ovens. Of the reportedly eight quality mines he purchased, several were located around the mining village of Anyuan while some were also in the eastern Gaokeng Mountains, sites near or directly controlled by the Ouyang family.[77] Lu complained, however, that this method did not work as well as he hoped because the local gentry—particularly the Wens—were using their hold over the county to hinder his attempts to buy up and take over local extraction. Specifically, he complained that Wen Tingjun started a rumor that led to conflicts between Lu and some of the men he hired.[78] Furthermore, Lu argued that when he bought up two-thirds of the shares of one mine, shares he understood to be adequate for control of the mines, the mountain lords argued that the shares he did not hold were worth more than those he did. In essence his shares were taken by the other miners as a minor partnership in the firm. He also stated that after he had agreed to buy one mine for 500 *yuan*, the owner increased the price to 620 *yuan*. Lu also complained that because he did not yet have working coking ovens, local producers gouged him when he purchased their products. Added to all these problems was the simple fact that a reported above-average rainfall that winter stopped all traffic out of the county and flooded some of the mines. Yet, with all these problems, Lu asserted, he could still ship more coal than the Guangtaifu Lineage Trust had been able to provide, and he further explained that by paying only 1 *liang* for shipment of every ton of coal, rather than more than 5 *liang* as the Wen lineage was paying, he kept his prices down.[79]

Even while Lu was making claims of economic efficiency, his tactics soon raised the ire of the local gentry families who had ruled the county for centuries. To the county's leaders, many of whom were mountain lords or were related to people who were, Lu must have appeared as an outsider of questionable status who used tactics that did not account for the pre-eminence of their own families and local positions. On March 22, 1897, twenty-one of Pingxiang County's most prominent gentry and lineage leaders collectively signed a petition that they sent to Sheng Xuanhuai listing Lu's faults and calling for immediate action. The petition begins by attacking Lu's assistant, Yang Shouquan, whom the gentry argue was not a native of Pingxiang County and thus should not have been a part of the project. First, Yang, they charge, deceived Lu when he said that the coal in northern Pingxiang County was superior to that controlled by the Guangtaifu firm and by not telling Lu that in fact powerful gentry leaders owned and operated Guangtaifu.[80] Second, the gentry complained that Lu bought up many mines at exorbitant rates, leaving him with a series of shafts that could not be easily linked together for modern expansion. Lu, they went on, then realized that the Guangtaifu Lineage Trust did have the best coal in the most accessible locations in the county, and so he began to take their holdings by force, thus violating agreements previously made between Guangtaifu and Sheng Xuanhuai. These actions, they argued, would certainly lead to other miners holding back on their coal to push the prices up, and this in turn would harm the county's economy. Third, the gentry were bitterly angry with Lu for establishing residence in the local Temple of War and living with "prostitutes" in the hall thus desecrating the county's ceremonial structure. Fourth, the gentry one last time criticized Yang Shouquan, calling him a bully who beat and imprisoned people who disagreed with him. The signatories concluded their letter by tactfully conceding that Lu Hongchang may have been deceived by Yang Shouquan, and they urged Sheng to remove Yang as Lu's assistant.

The petition then further explained that students were putting up wall posters in the county seat and elsewhere calling for Lu's removal. The signers ended the petition with a plea for assistance:

> Therefore, for one or two months we will all keep silent, but only because Mr. Lu mistakenly believed some rumors and recklessly took the above actions. This is not what he initially intended, but it has already become the situation. And so we ride on the tiger and take a risk. All these reactionary actions indeed caused damage to your bureau and our county. Reluctantly, we write this letter together . . .[81]

This letter is especially interesting for several reasons. First, the petition is signed by the "who's who" of the county. Li Youtang, the older brother of the scholar-turned-mountain-lord Li Youru, numbered among the signatories, as did two members of the Wen lineage and at least three men who were former Hanlin Academy members.[82] Second, the petition presents evidence of gentry distrust of the modernization scheme and the imperial government's direct intervention in their economy. Third, because they addressed the petition to Sheng Xuanhuai, a man who held the office of director general of the imperial railway administration and was Governor-General Zhang Zhidong's chief assistant, it indicates the degree of influence they believed they had over their county's internal affairs. Furthermore, the local leaders used their authority as gentry to exert a legitimate influence over local concerns and to try to rectify this matter. Rather than engaging in antigovernment activities or attempting to distance themselves from the actions of the Hanyang managers, they turned to official channels and followed the traditions of the Confucian order.[83] Also, while they directed their objections against the Chinese managers, they made no mention of the fact that, as is described in more detail later, the first German engineer was already in the county testing the coal deposits. This may indicate that the gentry accepted his role in the future of the county even as they became angered by the actions of several Chinese officials. Moreover, the document did not call for an end to Sheng Xuanhuai's plan to implement a major mining scheme in the county. It only argued that the initial attempts at expanding mineral extraction undercut the power of the local lineages and insulted their status in the local community.

Lu Hongchang replied to the petition with a statement of his own that clearly showed contempt for the local leaders. His letter was signed before the petition was sent but obviously responded to the gentry's letter. In it he complained that the local gentry greatly undermined his work—which he all but admitted included purposefully skirting around the Guangtaifu Lineage Trust—with their attempts to discredit him and his administration. He countered that his actions were indeed working. Specifically, while some of the most important of the Guangtaifu trust's mines collapsed from water damage due to rain, he purchased rich mines that Ouyang Bingrong and Yang Shouquan adroitly managed. However, he argued, his successes were not being fairly assessed by the local gentry, though he felt that in time his accomplishments would show themselves. The time would come, he argued, that "when the water subsides, the rocks will emerge."[84]

Even as Lu Hongchang was doing his bidding, Sheng Xuanhuai realized that Lu's efforts actually undermined his plans. To investigate this matter further, by July of 1897, Sheng sent an advisor named Li Zongjin to interview the people on the ground and submit a report. Li informed Sheng that the "Merchant Management" model of the Guangtaifu Lineage Trust was not efficient because the Wens employed the same brutal tactics with their workers as mountain lords of the past and ignored many of the quality-control strategies needed for success. On the other hand, Lu's attempt at "Government Management," though perhaps politically inept, was a more successful model that was better for the workers and the bottom line.[85] Based on Li's findings, Sheng decided that Lu Hongchang's overall strategy was the right one but that Lu himself was not the right man to take control of the coalmines. To this end, Sheng secured new sources of capital and established a more formal administration to oversee the entire county's mineral extraction.

A New Plan: Direct Control

Dedicated to the new scheme in Pingxiang County, sometime in 1898 Sheng Xuanhuai and Zhang Zhidong wrote a memorial to the throne requesting special dispensation regarding government investment in and control over the mines in Pingxiang County. They wrote that as they were attempting to make use of the mines, members of the local population were attempting to monopolize the coalmines in order to make a quick profit. They argued that these people did not understand the larger effort being undertaken and therefore had to be stopped. To this end, Zhang and Sheng called for a moratorium on all new mines and restriction on existing mine expansions as well as coal sales in the markets. The throne accepted these proposals and allowed Zhang Zhidong, the governor-general of Hunan and Hubei provinces, and Sheng Xuanhuai, the director general of the railway administration and of the Hanyang

Ironworks to take direct control of the coalmines in Pingxiang County, Jiangxi Province.[86]

Furthermore, the memorial announced that Zhang Zhidong and Sheng Xuanhuai sent one Zhang Zanchen to Pingxiang County to act as supervisor of the mines. The two men believed that he was a skilled businesses manager, and he was a wealthy merchant and a gentry member who held the rank suitable for county magistrate-in-waiting by purchase. Moreover, he was born in Wujin County in Jiangsu Province, and therefore he and Sheng Xuanhuai were *tongxiang*, or people from the same county.[87] While Zhang Zanchen was initially hired by Sheng Xuanhuai to be a manager in his Hanyang factories, Sheng decided to send him to Pingxiang County as part of a much more intense effort at taking direct control over the county's production scheme.[88] Though he apparently did not have the expertise in chemistry or mineralogy that Lu Hongchang had, Zhang possessed other essential abilities. The memorial stated in part:

> We have investigated and found that there is one Hubei County magistrate-in-waiting named Zhang Zanchen. He has an upright character and is respectful and stern. With important affairs he knows what is right and sticks to it. When things need to be repaired, he is exacting. He has the necessary talent to meet the challenge.[89]

The memorial describes Zhang as conscientious and dedicated. Another depiction of him indicates that he was "sagacious" in his dealings with his workers. But none of the documents indicate that he was conversant in engineering or that he understood the tasks he was being called upon to oversee.[90] Rather, Zhang was to supervise the modernization process and to make sure the transitions went smoothly.

In their request Zhang Zhidong and Sheng Xuanhuai also provide a general overview of what was planned for the minefields, pointing out that the territory around the mine was about to be transformed by a "great undertaking" (*da ju*) that included "arrangement of a vast scheme . . . building roads and establishing telegraph lines, transporting machinery," as well as working with Chinese and foreigners alike.[91] Zhang Zanchen's duties were even more expansive than that, as soon after he arrived he was called upon to push aside gentry mountain lords who commanded the premodern mines and replace them with foreign engineers overseeing mechanized mines that were several times the size of the indigenous shafts. Furthermore, he was to assist in the construction of a railroad that was designed to replace local boat haulers and transportation workers who had hauled goods into Hunan Province for generations. In sum, Zhang Zanchen had to take control of the economy of a county he had never lived in before. He was not handed the bureaucratic portfolio of a magistrate or member of the centuries-old bureaucracy. In fact, he was not a member of the literati as much as he was a merchant with a purchased degree. Rather, he was acting as the deputy of the newly formed imperial railroad commission, whose administrator was himself as much a part of the world of China's most powerful merchants as he was a high official of the court. While this position in the official bureaucracy slightly echoed that of the salt monopoly of centuries past, his responsibilities encompassed far more expansive and ultimately transformative duties.

Assuming the positive response from the court, Sheng Xuanhuai sent Zhang Zanchen to the Pingxiang County Coalmines in the spring of 1897 nearly a year before the memorial reached the court. By August of that year Zhang had arrived and taken command of the modernization project, thus officially making him Sheng's manager over the county's mine bureau. In a letter to Sheng, Zhang wrote of his initial actions and assessments regarding the condition of the mines and the leadership in place. First, upon arriving at Pingxiang County, he told all the officials to send as much coal downstream as they could. Because there had been accusations of hoarding, it is likely that Zhang felt the need to address this problem immediately. He further wrote in his first letter that he examined the mines' ledgers right away to find any and all corruption regarding extraction and distribution. Second, he began to actively buy up as many mines as he could afford.[92] This was a difficult process, he complained in the letter, because the mines were scattered throughout the mountains and miners hoisted countless flags by each shaft in order to stake their own claims. The flags, he wrote, were "as numerous as a forest of masts" and the mines were "as densely packed as a beehive."[93] This was a problem for all of the previous managers in part because they hindered central control over production prices and technology and because the smaller, less modern shafts caused the larger mines to deteriorate and flood.

Zhang also provided assessments of the men in charge upon his arrival. In particular, Zhang wrote that Lu Hongchang's actions were largely correct, but his boastful personality and rash and impetuous style almost certainly at least partially caused all the problems in Pingxiang County. More generally, Zhang argued that while some of the firms employed by Lu and his predecessors were good and effective, some were not. Zhang proposed again to separate the good and bad mines and to eliminate the problem firms. He planned to fire the officials whom he deemed troublemakers. Similarly, Zhang set out to find officials responsible for various acts of corruption and remove those men not working in the interests of the project.[94]

While Zhang Zanchen was reordering the mining scheme in Pingxiang County, he and Sheng Xuanhuai sought the financial means to integrate the technology necessary for a first-rate coalmining scheme. First, the two men worked together in Hankou and Shanghai to raise investment capital from local merchants, and in 1899 Zhang founded a bank in Pingxiang County.[95] Then, by 1902, Sheng secured foreign loans and Chinese official funds including 1.5 million taels (about 4 million marks) from the German firm of Carlowitz and Company and over a million taels from large Chinese firms including the China Merchants' Steam Navigation Company as well as small banks and private investors.[96]

Under most industrial loan agreements between countries, the bank that made the loan either appointed an engineer for the project or required that the engineer be from the same country as the lending bank. Since the first large foreign loan for the industrial development of the Pingxiang County Coalmines came from Germany, the chief mine engineers were likely to be German as well.[97] Even before Sheng secured the loan, a German engineer named Mr. Marx was working as the administrator of the Daye County mine engineers (*zong kuangshi*) charged with the task of modernizing the mines for Zhang's early industrial efforts. In fact, Marx was so impressed with the area's iron ore deposits he attempted to get the German government to pry the minerals from China's grip.[98] Once Sheng Xuanhuai took control of the Hanyang Ironworks and its subsidiary mines and firms, he hoped that Marx would act as the head engineer in the coalmines in Pingxiang County. To this end, in a letter to Sheng Xuanhuai dated June 24, 1896, Sheng's assistant, Zheng Guanying, ordered Marx to go to Pingxiang County to investigate the mines there.[99] Two days later, Yun Jixun, one of Sheng's subordinates, called on Sheng to send Marx to the county as soon as possible.[100] However, Marx refused to go to the Jiangxi highlands. This refusal, Zheng argued, was in violation of his contract, though he did not suggest any punishment for the insubordination. Being unable to compel Marx to complete his duties, Zhang instead suggested to Sheng that they send an alternative representative.

The logical choice for this position fell to Marx's trusted subordinate and fellow German engineer, Gustav Leinung. Leinung complied with the command, and in the fall of 1896 he set about to travel from the

Daye area in eastern Hubei Province into the northwestern-most cities of Jiangxi Province. When he reached the river-port town of Jiujiang located south of the iron ore mines in Daye County in the late summer of 1896, he was met by Yun Jixun. Since Yun sent a letter requesting that Sheng send his best German engineer to Pingxiang County, he no doubt was excited to greet an equally fine engineer willing to oversee the coalmines. After they met, Yun escorted Leinung to Pingxiang City. Leinung had to be escorted by a Chinese official in part due to the legendary, yet all too real, xenophobia of the local population. When the contingent reached the eastern outskirts of the county's border, they learned that Leinung was in actual physical danger. Miners in the town of Luxi who opposed the coming of the foreign engineer had gathered along mountain ridges above the route and planned to ambush him with a hail of rocks when he passed by.[101]

Fortunately for Leinung, Yun Jixun was able to make alternative plans, and the German arrived safely in the county seat. However, even after Leinung entered the city walls the county government provided military protection and by various means sought to accustom the people to the foreigner's presence and calm the simmering unrest among them. In an article in the *The Far Eastern Review* based on an interview with Leinung, one author wrote:

> Upon his first arrival in [Pingxiang County] he was an object of awe and amazement to the natives who had never seen a European. All sorts of beliefs were held about him. He was said to have three eyes, one at the back of his head, and it was believed that he could see deep into the bowels of the earth and detect the treasures there. When he entered [Pingxiang City] for the first time in 1896 the natives were sitting on the roofs of houses to see him. He was put into a small room in an ancestral hall in which there was a grated window, and the people were allowed to come along in squads of 10 or 12 to have a look at him, as though he were a rare zoological specimen. The idea was to make them accustomed to the appearance of a foreigner. During the first year there were always 200 soldiers around Mr. Leinung. He was not allowed to leave his quarters and had strictly to obey orders.[102]

This depiction of the early months suggests a very tense period throughout the county. Normally, Pingxiang County's local militia forces were estimated at about 250 men and therefore a contingent of 200 troops strictly to protect one man is very large indeed.[103] At the same time, Leinung's apparent acquiescence in being turned into an exotic creature for the local population indicates his own awareness of the situation and his diplomatic skills in trying to curb possible tensions and violence due to his presence. In fact, while it was certainly true that many people thought the foreigner had special powers and even devilish plans for the people of the county, it is also true that the region's most educated young men understood the potential consequences of foreign penetration into the markets and resources of their county, since newspapers and bulletins routinely published emotional accounts of wars and conflicts of the previous decades.[104]

For the next several months Leinung tested the mines in the Jiangxi provincial highlands. While he apparently focused his studies in the region around the town of Anyuan and the holdings of the Wen lineage, he reported that he also examined deposits in other parts of the county and even in the mines in Yichun County to the east.[105] His findings indicated that the coal throughout the area met his quality standards for the production of steel and pig iron. In fact, Leinung once explained to a Pingxiang County student that the deposits were about ten miles long (30 *li*), three miles wide (10 *li*), and 48 feet deep (48 *chi*) and were second only to England's in overall quality.[106] He went on to tell one Western journalist that the coalfields would last "several hundred years with an output of 1,000,000 tons annually." Once washed, the coal was shiny and contained no more than 10 percent ash and less than 0.05 percent sulfur.[107]

Based on these tests, Leinung began to put together a plan to combine many of the small mines together into one large mining scheme. He envisioned a large horizontal adit, connected by an electric train that would begin near the town of Anyuan and would traverse several mountains and the mines they contained. Eventually, he hoped to construct a second adit that would travel slightly to the west of the first but that would also terminate in Anyuan. To bring about this project, in the spring of 1897 Leinung informed the new mine bureau director, Zhang Zanchen, of the mines and properties he wanted to integrate into the mechanized mining scheme.

Leinung continued to do surveys of the region until finally, during the middle of June of 1903, he wrote a long letter to Sheng Xuanhuai delineating his findings and the budgetary issues regarding the expansion of either Daye County or Pingxiang County as the centerpieces of Zhang's industrial empire. He first explained that Daye County held more iron ore than nearly any European country and that only Sweden and Norway contained more iron ore.[108] Mining officials in the area placed the estimates into the millions of tons. In fact, Leinung is said to have put the number as high as 103 million tons.[109] He did advise Sheng that the iron ore was not as pure as that found in Europe as it contained .8 percent phosphorous, while most hematite iron ore should contain no more than .5 percent.[110] However, overall, he believed Daye's iron ore sufficient for the task at hand.

Second, Leinung argued that Pingxiang County's coal deposits could provide the main source of bituminous coal for smelting the iron ore being mined in Daye County. He did mention to Sheng Xuanhuai that the Pingxiang County Coalmines were slightly high in ash, an issue that had to be dealt with in the purification process. Mention was also made of the iron ore mines in Pingxiang County. Leinung also deemed these smaller iron ore fields in the Shangzhu Mountain range in the westernmost reaches of the county adequate for the making of iron and steel by the firm.[111] Finally, though Pingxiang County was located in the isolated highlands between Hunan and Jiangxi provinces, Leinung assured the Chinese that transportation costs would be resolved with the construction of a railroad from the mines to the ironworks.

According to the local gazetteer, when Lu Hongchang returned from his first investigation of the county's coal, he informed Sheng Xuanhuai and Zhang Zhidong that the coal deposits were of high quality and sufficient quantity to solve the ironworks' needs. In an emotional description the author explained what happened next:

> Lu Hongchang returned to Hubei and gave a report on the situation of the (Pingxiang) mines. The materials were rich and beautiful and could be compared to those in Kaiping. Governor-General Zhang Zhidong and Vice Minister Sheng Xuanhuai jumped to their feet when they heard this! They sent a memorial to the throne requesting that the mines be opened.[112]

This quote, which describes an event likely taking place around the last months of 1896 or the beginning of the following year, portrays the culmination of a series of steps and missteps instigated by Zhang Zhidong and Sheng Xuanhuai that involved integrating Chinese bureaucratic orders with Western technology to discover and extract coal from a suitable central Chinese minefield. By the summer of 1897, Sheng and his men chose Pingxiang County to fulfill that duty. Quickly, a small cluster of outsiders, foreigners, and local dignitaries arrived who all hoped to bring about the new economic order in the county. Among the local inhabitants, the Wen lineage leaders were residents of Pingxiang County and had thus been part of the county's political scene for generations. To get what they needed out of any new imperial policy, they had to deal with fellow lineage members, competing lineage leaders, and even the county magistrate. They viewed Sheng's plans within the context of their patron-clientage relations but also viewed them as a possible venture for breaking out of the systems of the past in the hopes of dramatically increasing their wealth and status with the powerful leaders of the imperial government. Unlike the Wens, Zhang Zanchen and Lu Hongchang were Chinese merchants and semi-skilled engineers hired as part of a Confucian-based bureaucracy that was stretching its abilities as far as possible to meet up with the new world order. As with the emerging managerial class in the United States these men used their Confucian-based education and hands-on training to transform local patron-client systems of production into efficient schemes suitable for Western machinery. Finally, Gustav Leinung was called upon to use his foreign engineering background to alter China's physical landscape while revolutionizing the social and managerial structures of production. His hopes for Pingxiang County will become clearer in the following chapters, but suffice it to say that he probably saw himself as someone who could do a great deal of good for the Chinese people, and he agreed to do that for a fee.

What is especially significant, at least for the Chinese people involved in this project, is that Pingxiang County's mines and Sheng Xuanhuai's administration provided new avenues for social and political advancement. Even as court officials planned to develop a coalmine in an isolated community located in the highlands of southern China, they devised the scheme as part of the court bureaucracy. To facilitate the court's needs, government officials hired men who were either low-level literati or officials-in-waiting as managers and overseers and granted them official bureaucratic grades they were hoping for. Moreover, merchants who in the past would have seen no hope of achieving official rank now turned to the newly emerging branches of the court—such as those that oversaw railroads, telegraph lines, banks, and armories—where they could hold prestigious posts and become part of the largest firms in the empire. For China's government, these men were part of the "self-strengthening" that the empire so desperately needed. For these men, the government's new policies offered a chance of promotion into the emerging world economy that promised wealth and status.

However, just as the men who succeeded in the new court bureaus came from different social and geographic places, so too did success promise different rewards and compensation. For local gentry leaders like the Wens, industrialization in their home community acted as a carrot dangling in front of them and promising to further entrench and expand their local power while pushing their economic connections outward. These men began to envision the chance to get away from the countryside and become part of the urban elite while leaving their peasant brethren struggling with the "crisis of authority" that was one of the most damaging features of global modernization.[113] For the outsiders, on the other hand, Pingxiang County itself held no personal importance; it was a means to an end. It was a blank slate, a mineral field that had been improperly handled in the past and needed a complete makeover for future success. For them, local concerns were not relevant to their tasks ahead, and success did not come by seeking the best life for the community but by accomplishing the goals of the court by whatever means possible. Since the goals of the project at different times pitted the local community against outsiders, elites against nonelites, and Chinese against foreigners, the means to each end often required alliances of any number of these people against the rest. With so many different allegiances, each of the players approached the remapping of the county with their own interests, some of which were complementary and some of which were not. Winners and losers, allies and enemies, emerged, formed, and reformed as each new layer of change took over the landscape. In the last chapters I discuss these important changes and show how the hope of economic progress altered the notion of community and the definition of success in Pingxiang County.

4

Irrevocably Remapping the County

The rapid and dramatic reorganization of the steps of industrialized production and consumption can have social and cultural ramifications not discussed in most Western studies of the modernization process. In China, the European imperialist powers "deterritorialized" the empire, that is, they broke down those aspects of the culture that resisted or were ill-suited to imperialist market forces. Conversely, once they reordered the sociocultural structures in place to facilitate plunder, they then "reterritorialized" the landscape so they could establish permanent and reliable forms of extraction.[1] With these changes, imperialists imposed new lines of communication and bureaucracy on top of old ones, forcing Chinese commoners and officials alike to adhere to a new world order. In some cases, the Chinese people negotiated and resisted certain aspects of this reconfiguration. They pushed Western engineers to accept some features of Chinese production methods and labor styles and even organized space and time to fit with local culture. However, the Westerners held ultimate authority over the efficacy of the new order and they made clear that certain aspects of the scheme were nonnegotiable.[2]

In particular, the foreign engineers demanded that Chinese officials redraw the local landscape to facilitate mining and transportation. Agreeing to these plans, they purchased or took control of farm fields, villages, cemeteries, mines, and factories that had been the centerpieces of subsistence livelihoods for generations. Then they reorganized those properties in order to provide for the industrial scheme. Paddy fields and vegetable gardens became new points of industrial production, and river routes and overland trails lost their importance to newly constructed railroad tracks and vehicle-useable roads. The new roads in turn linked towns together in new ways. Some river ports lost their importance, while other villages located in open valleys drew train traffic and supporting merchants to their stations. These economic changes had serious social connotations.

Lineage leaders and wealthy landlords who relied upon their fields for rents handed over their properties to government officials for use in the new economy. This meant that not only did they lose their land, but that they had to sever their economic connections with poorer lineage members and subsistence peasant laborers. Peasants who farmed particular fields for generations had to give up their access to the landlord's properties and seek out new patrons. Those who worked as boat haulers and land-based transportation workers lost their jobs to the railroads, while underemployed miners and factory workers gained employment in the foreigner's factory or searched for a job elsewhere.

At the same time, even as the foreign engineers demanded local compliance, many local populations negotiated and altered specific features of the industrial scheme in order to maintain some aspects of their culture and lifestyles. Local officials charged with acquiring properties to satisfy the Western engineers had to purchase lands from leaders within the community who were often essential to the county administration's success at maintaining peace and prosperity. When county officials attempted to take landlords' properties, the local elites sometimes used their considerable leverage to negotiate favorable purchase terms. Similarly, poorer families forced to give up their properties negotiated with officials to pay them well to assure their collective survival, which the government also demanded. Therefore, even as imperialists and their Chinese allies sought to redesign China, they were at times forced to do so within the confines of Chinese culture and social norms. This give and take altered the production process in the Pingxiang County Coalmines and in turn created new and unique issues of production.

Taking the People's Livelihood: Purchasing the Mines

A great deal has been written about land purchases and contracts in Qing dynasty China. Most of these studies examine the informal and formal means by which people acquired properties for their personal use as opposed to the official acquisitions undertaken here.[3] Using local land contracts and the memoirs of Pingxiang County's magistrate, I show that the two government institutions—the county magistrate's *yamen* and the Mine Bureau under the leadership of the Imperial Railroad Administration—forced the local population to sell their lands employing a form of eminent domain.[4] In China, the emperor claimed all property of the empire as part of his personal domain. However, Confucian dictates required that the emperor be benevolent toward his subjects, and in this manner he conferred the lands of the throne upon the people for their own use. In those cases when the throne did decide it needed to take possession of empire lands, it was the responsibility of the government to make sure that such actions did not harm the population. In other words, as government officials like Sheng Xuanhuai located lands they wished to expropriate, they needed to be aware of the concerns of the local population and use their powers wisely so as not to cause unrest.

In the case of Pingxiang County, in 1898, after Chinese experts and foreign engineers tested the properties, Zhang Zhidong and Sheng Xuanhuai sent a memorial, or an official letter to the throne, requesting the rights to extraction and distribution of the county's coal. Citing the precedence of the Kaiping Coalmines, Zhang and Sheng requested that they be allowed to purchase the lands suitable for modern mining in the area and that they be given the rights to stop all others from mining in the area in efforts to compete or otherwise hinder in this scheme. The court granted this request, giving the two men the authority to purchase minerals from local producers and to engage in various takings of local property in the name of the court, as long as they paid a rate for the land as explained by previous self-strengthening precedent.[5]

Sheng subsequently sent Zhang Zanchen to Pingxiang County in June of 1897 and commanded him to find the best means available to securing coal for use in the Hanyang Ironworks. Zhang's investigations found that the Guangtaifu Lineage Trust remained the largest single mining firm in the county. Led by Wen Tingshi and Wen Tingjun as well as other members of the Wen lineage and several non-Wen gentry figures, Guangtaifu controlled several of the county's best mines, owned its most sophisticated equipment, and developed the strongest political ties both within the county as well as with outside figures such as Zhang Zhidong and Sheng Xuanhuai. In fact, Sheng, aware of the power and stature of the Wen lineage in the area and the significance such men played in the county, argued that getting rid of them was immoral and counter to Chinese values.[6] However, while Guangtaifu had successfully been able to send a few tons of coal to Hubei Province, the lineage trust was still nothing more than a premodern mining firm trying to satisfy the needs of a modern ironworks. Under these circumstances, Sheng Xuanhuai and Zhang Zanchen knew the Wens' stranglehold on the county's mines had to be broken and that some plan—hopefully one beneficial to both the Guangtaifu Lineage Trust and the Hanyang Ironworks—had to be implemented.

These initial hopes for a suitable compromise quickly soured when, during the summer and fall of 1897, the Guangtaifu-owned mines suffered several accidents that cost the Wen Lineage Trust money and made it impossible to deliver sufficient quantities of coal to the ironworks in Hanyang. Guangtaifu's problems were both troubling and fortuitous for Zhang Zhidong and Sheng Xuanhuai, as they provided the Hanyang leadership with a legitimate excuse to take over the Wen lineage mining operation.[7] Seizing the opportunity he was looking for, Zhang Zanchen coerced the embattled Guangtaifu Lineage Trust to sell its entire operation—including, by now, seven Western-style coking ovens and eighteen mines—to the Mine Bureau sometime during the last months of 1897.[8] With this purchase, the Wen leaders, some of the county's most powerful lineage leaders and landholders, were soundly defeated by government pressure and left incapable or unwilling to put forth even token resistance.

From this point on, they decided to walk away from the Mine Bureau and the mechanization project. In particular, Wen Tingjun's role in the mechanization of the mines cannot be found in the records after this date. His name does not appear in the land contracts located in the Mine Bureau archive, nor is it mentioned in the Sheng Xuanhuai archival correspondence or in any of the subsequent studies of the mechanized mine. As will be discussed in chapter 6, Wen Tingshi does reappear but this time as an opponent of the Mine Bureau's desire to take control of the Shangzhu Mountain iron ore deposits located near his family's settlement.

With this purchase, the Guangtaifu Lineage Trust, the county's most prosperous coalmining firm, was simply bought up and integrated into the Mine Bureau along with many other mines around the village of Anyuan. The Wen lineage gave up their properties that promised financial success and that had been so important in the early months of the modernization process. Beyond the firm itself, the Wens lost several of the county's most lucrative mines and the adjoining lands. Seasonal laborers used the mines in exchange for rents, and the quality and quantity of the deposits promised to pay in this manner for generations to come. Finally, and perhaps most importantly, by giving up the Guangtaifu-held properties the Wens lost the patronage of the seasonal laborers who worked the mines. Once Zhang Zanchen possessed the mines, these men had no alternative but to take coalmining jobs from the Mine Bureau or, alternatively, from another mountain lord. Since coalmining provided both an income for the family and fuel for the home, some of the men who worked the Guangtaifu mines severed their ties with the Wens and went elsewhere.

Zhang's purchase of the Guangtaifu properties forced the Wens to accept the loss of both labor income and the tenants who provided it.

Having now taken over the single largest gentry-held mining consortia, Zhang Zanchen, with 510,000 taels of silver borrowed by Sheng Xuanhuai, systematically purchased the remaining lands that the German engineer, Gustav Leinung, desired.[9] These purchases included active mines and the adjoining irrigated fields and gravesites of the peoples' ancestors Leinung deemed necessary or potentially lucrative. To help him in this endeavor, in March of 1898, Zhang established a new Mine Bureau with another Jiangsu provincial native determined to modernize China, one Li Shouquan.[10] Given the archival collection of land contracts, it is certain that Zhang's Mine Bureau purchased hundreds of pieces of land of various sizes and values. In a brief history of his actions in the early years of the mine, he explained his success at purchasing the lands while at the same time being aware of the needs of the local population:

> In succession [I] bought and attained fields and mountains of more than 1,700 *mu*. . . . The Pingxiang [County] commoners view working in the coalmines as their livelihood. . . . Because we have already received commands, I will not allow the opening of many small mines. Also, a priority is to buy out each of the merchants who [have mines that] border on Anyuan and consider granting them sufficient payment in return. [In this way] we take the people's livelihood and our own obedience to the government's regulations to be extremely important.[11]

Based on this statement, it is clear that Zhang assumed the right to purchase all the lands around Anyuan but that to do so he needed to enter into proper agreements with each of the local landholders. That is, he did not feel he had the right to simply declare all the property to be under the protection of the empire and indiscriminately force all the people out. Zhang also tells us that after he purchased the shafts, he hindered further mining in the lands directly adjacent to the shafts he now owned. This was in part to stop the further manipulation of mining and coal sales in the region and in part to stop poorly constructed mines from flooding the mines he was going to run. Furthermore, due to the inexact determination of the borders of each of the claims around the shafts, Zhang could not be certain that he actually held all the lands or that the land contracts he held would prevent others from starting up new shafts nearby. Conversely the vagueness of the titles he signed likely gave him some room to expand his own claims when the situation was required.

The Mine Bureau contracts I collected from an archive in the county were written between the year of Zhang's initial purchases in 1898 and 1900.[12] These documents indicate that the bureau gradually purchased lands piecemeal to take more complete control of the region. Some of these contracts portray the imprecision described by Zhang's troubles. One was for property at Zijiachong being sold by Wen Pinshan. The contract stated that the property was three mineshafts, though it did not indicate the amount of land in *mu* or any other form of calculation.[13] Similarly, the Mine Bureau purchased a piece of mining property located in a bamboo grove near the Gaokeng Coalmines, but the contract gave no exact boundaries for the property.[14] Several other purchase agreements describe irrigated lands providing explanations of the boundaries of the fields but then adding that the property also included one or more "hills" that were not further defined. One contract, for example, states that the property to be purchased included twenty-one hills.[15]

Zhang's primary purchases centered around the village of Anyuan, the area where most of the best mines were located. This location, rather than the many towns and villages along the rivers that held coal markets in the past, was to be the focus of the mining scheme. Specifically, Leinung planned to dig adits—that is, horizontal caves that followed the mineral seams—into the highlands around Tianzi Mountain to join the vertical mines following the most lucrative deposits. In his first purchases, Zhang bought and began using some fourteen shafts around the village of Anyuan, while he boarded up and closed sixteen of the poorer mines in the vicinity. Zhang then established concentric circles of control for those mines located beyond this central location. Those lands deemed beyond the direct control of the mine bureau remained under the ownership and control of the local mountain lords. He deputed Lu Hongchang, the official who was criticized by the local gentry for his brutal tactics trying to skirt Guangtaifu, to organize these mines into "partnerships" (*bao he gong zhuang*). Under this arrangement, Lu purchased the extracted mineral from the mine owners to supplement the initial output of the Anyuan mines. Zhang and Lu completed this work with the help of gentry leader Li Jingshu, who used his influence as a local leader and a Japanese-educated scholar to protect local miners from the oppression they feared they would suffer under Lu.[16] Mines controlled by the partnerships could also sell their coal to local firms that in turn delivered the mineral to the county's markets and homes. Finally, mines located even farther from the immediate circle that were too small to provide for the needs of the bureau continued to sell their product in the local markets including, presumably, the Mine Bureau state market at Anyuan.[17]

Even as Zhang Zanchen simply explained in his brief history discussed earlier that he properly purchased mine property for the state, in reality his purchases broke apart and reordered significant aspects of Pingxiang County's society and economy. The Wen lineage that had been the single most powerful coalmining firm in the county was subsumed into the offices of the state. Privately held coalmines that were central to subsistence and patronage strategies for decades if not centuries were now part of the new government-run firm that from that time forward owned all the land, managed all the plants, and held all the jobs. And even the centuries-old coal markets located along the river ways and land passageways lost their importance as the coal that was mined from throughout much of the county was delivered to a central location in the village of Anyuan. Once the mineral reached the holdings of the mine bureau in Anyuan, it was to be distributed beyond the confines of the county borders first using the river routes and then, eventually, employing a railroad branch line. As will be shown in the remainder of this chapter, to construct this railroad line and replace the trade routes of the county's past even more land was purchased and reorganized, and, as with the mine purchases, the social and economic consequences were significant.

Trains, Schools, and *Fengshui*: Purchasing the Railroad Property

Along with the accumulation of mines and coalfields, Sheng Xuanhuai planned to replace the inefficient boat haulers with a modern railroad line that could transport coal from the mines to the Wuhan Cities. To this end, he needed to purchase land from the mines in Pingxiang County to the Xiang River that could provide for a railroad branch line. This included linking this branch line to the trunk line he was constructing between Guangzhou and Beijing with a stopping off point in the Wuhan Cities and the Hanyang Ironworks. In 1898, in order to fund this endeavor, he used capital from the railway administration to strike a deal with the American-China Development Company, the company that his office already hired to assist in the construction of the southern half of the trunk line to the Wuhan Cities. The company was made up of well-connected Americans, including William Barclay Parsons, a railroad engineer who was a central figure in the construction of the New York subway system, and his investors, who included one former United States senator and several prominent merchants from the United States.[18] Under the agreement, the American-China Development Company provided a loan of 20 million dollars for the initial costs, including the purchase of land that was to be contracted for at market prices.[19]

With the agreement in hand, in the winter of 1898 and 1899 Parsons traveled along the proposed rail line with an entourage that included W. W. Rich, who was to be the chief overseer of the Pingxiang-Zhuzhou, or Ping-Zhu, branch line.[20] Rich, whose corrupted Chinese name, Bai Lizhi, suggests that his given name was William, or "Bill," is referred to in legal documents as "the Engineer of the Railway Administration" and as the "Consulting Engineer to H. E. (His Excellency) Sheng."[21] Unfortunately, not much more is known about Rich, though apparently, like Parsons, he was an American engineer with a background in railroad construction.

The entourage traveled from county to county making plans about how the line was to be constructed. When they reached central Hunan Province in early January of 1899, they headed east and traveled along the Lu River up the Luoxiao Mountains to Pingxiang County where, as was customary for all foreigners entering a county, they met the county magistrate, Gu Jiaxiang. Rich and Gu then quickly drafted a proposed line that began at the town of Anyuan and proceeded northwest to the county seat of Pingxiang City. From there, the rail route twisted and turned westward to the town of Xiangdong, an important market town and the home of the Wen lineage.[22] It then continued past the provincial and county border and reached the county seat of Liling City in Hunan Province. Eventually, the line was to continue in a northwesterly diagonal until it entered the budding artisanal city of Zhuzhou, where it was to meet with the trunk line that followed the Xiang River into Hubei Province.[23]

Once the line was platted out by Rich, the American left the county and Magistrate Gu Jiaxiang was ordered to begin systematically purchasing all the necessary properties before Rich's return. Magistrate Gu came from Shaoxing County, Zhejiang Province, located in one of the regions of China most affected by Western economic growth. Specifically, the county was located between the important financial center of Ningbo and Hangzhou, the provincial capital and tourist location for the rich and powerful. Shaoxing was also noted for its successful literati who were posted in offices throughout the empire. Therefore, serving as the magistrate of the peripheral highland community, Gu brought his considerable political connections and a personal understanding of China's emerging growth and relations with Western countries and corporations. He initially served a term as the magistrate of Pingxiang County from 1888–1894 and then left to take a more prestigious post. However, when his replacement made a series of errors in judgment, the local gentry were unhappy with the new magistrate's performance and requested that Gu be reappointed as the county's magistrate. Gu Jiaxiang's second appointment, which lasted from 1896 to 1902, therefore coincides with the period of the establishment of the mechanized coalmines and the railroad land purchases.[24] In 1898, less than two years into his second term, Sheng Xuanhuai ordered Gu to establish a Land Purchasing Bureau and begin the process of securing the properties they would need.[25] Given his understanding of China's modernizing economy and society back in his homeland and his background as a successful and dutiful official, Gu complied.

There are no detailed accounts of all the lands secured by the county magistrate, but Gu wrote a memoir of the problems he faced and the strategies he used to resolve them. In it, he explained that he, along with several members of the gentry, laid out the route based on Rich's initial demands. They stuck flags into the ground along the pathway and separated all the lands that he had to purchase as part of the railroad route from those fields that were to be left alone by the state and allowed to continue as farm fields.[26] Gu was directed to buy parcels of valley property that essentially followed the Lu River from the mining area of Anyuan down to the county's border. These lands were often among the county's most lucrative rice paddy fields owned by some of the county's wealthiest landlords. But given Sheng's orders, Gu was charged with forcing the county's most illustrious leaders into land sales they did not want. At a deeper level, these eminent domain purchases represented a tug-o-war between the county yamen's claim as the court's appointed ambassadors and the local gentry whose status was based on their fulfillment of the Confucian social order and raw economic power.

One of the initial conflicts between County Magistrate Gu and the local gentry revolved around the need to construct the rail line across the Ao Island located in the middle of the Ping River that acted as a moat around three sides of Pingxiang City. The Ao Island is perhaps 400 yards long and 100 yards wide. The eastern half rose several feet above the river and acted as a breakwater, while the western half was lower and contained sandy soils but included some croplands among the marshes. The island was located just outside of the county seat's Lesser Western Gate where W. W. Rich planned to establish a train station to supply the nearby markets and provide a convenient location for passengers. Rich apparently further decided that the island would provide solid ground for a bridge support beam where the rails crossed over the river away from the city walls.

However, this island contained one of the most prized possessions of the local gentry, Ao Island Academy (*Aozhou shuyuan*). The county officials and local gentry had turned the island into a center for their religious and academic activities that made it among the most important centers of the county. Dating back to the late sixteenth and early seventeenth centuries, the island's east end contained a temple complex named Ao Pavilion that included a small Wenchang Temple where gentry prayed for good luck on their civil service exams. In 1709 the local magistrate constructed a Buddhist temple near Ao Pavilion on the island's eastern point facing the city. Finally, in 1756, near the Buddhist temple, the magistrate established Ao Island Academy.[27] During his first tenure as the magistrate, Gu Jiaxiang went to Ao Academy every month and tested the young scholars in preparation for their civil service examinations. Upon his return as magistrate for his second term beginning in 1896, Gu enacted progressive education strategies on the island, including abandoning the classical eight-legged essays in favor of more modern forms of testing. Gu remained steadfast in his interest in teaching the young students of Pingxiang County, even as the responsibilities regarding the establishment of the railroad placed greater and greater burdens on his time.[28]

The importance of the island in the eyes of the local population can be seen also by the fact that the name "Ao" is the first character in one of the alternative names for the prestigious Hanlin Academy as well as a title for the person who finished first in the examinations into the academy. Therefore, in naming the local academy with this title, the founders were placing great emphasis on its power to generate scholars of merit.[29] The 1935 edition of the local gazetteer discusses Ao Academy in glowing terms and pointed to its vast library as a point of special pride.[30] Among the maps in the gazetteer is a fish-eyed-lens-like drawing of the panorama of the county seat and the Ruyuan Pagoda and walkways outside the city that one would see while sitting along the island banks reading poetry or studying for the civil service exams.[31] The island's name also signifies a sea turtle that holds up the earth, and thus it was viewed in *fengshui* terms as a landmass that protected the city and required special care by its inhabitants.[32] The island and its academy were so important to the people of the city, in fact, that to this day it is still named Ao Island Academy even though the school and religious structures are no longer extant.

As early as 1898, after Sheng Xuanhuai called for the creation of a railroad line through the county but before Rich arrived to map out the route, Gu had arranged to purchase six *mu* of temple lands on the island. To this end, in January 1900, Gu asked one *jinshi* holder, eight *juren*, and thirteen lesser literati including Wen Tingbi, a member of the Wen lineage, to oversee the transfer of all lands required for the railroad.[33] He warned them to not criticize his decision to use Ao Island property. Instead, he demanded they clear the marshy areas and build up the lowest lands of the island to prepare a swath of land 23 *zhang* wide (about 270 feet) for the rails.[34]

Furthermore, Gu secured assistance of temple officials and leaders of other institutions whose lands he planned to purchase.[35] The temple officials did not want to sell this property and so Gu suggested a swap in which the land on the island be traded for tracts held by the imperial government elsewhere in the county. However, the gentry were still not satisfied with these terms. In one letter signed by several gentry members, the authors argued that the western half of the island served as the extended fields behind the academy and was unacceptable as a location for the rail line. They therefore opposed the purchase or transfer of the island property in any shape or form. Moreover, they explained that these purchases were destroying the local *fengshui* and that Ao Island was an important piece of territory for the continued strength of the county.[36]

Gu responded vigorously to these complaints, stating in part that the local gentry's arguments were without merit. He emphasized that the needs of the county were not as important as the needs of the empire. First, in order to hold off the Western powers, Pingxiang County had to develop even if that meant enacting policies that discomforted the local population. In this case, Ao Academy lands were as acceptable for purchase as any other. Second, Gu explained in a brief history of the county that lands were frequently transferred between owners. In fact, in several situations, the lands held by one person or one institution or another came under scrutiny due to poor bookkeeping by previous magistrates. Gu explained that under these circumstances lands were switched at the behest of the magistrate and with the acceptance of the landholders themselves. Thus, historical facts disproved any notion that land could not be transferred.[37]

Finally, the issue of *fengshui* continued to be a difficult problem for purchases. From the first railway built by the British, Chinese argued that the steel rails greatly harmed the *fengshui*. The American engineer William Barclay Parsons recalled a case strikingly similar to the Ao Island controversy of a railroad being constructed over an island close to a county seat that was described by the local gentry as possessing important *fengshui* attributes. In two different sources, Parsons alternately explained that the island was either said to be a large fish or a dragon that could not be cut in half by the railroad lest it would die.[38] However, Gu quoted a previous magistrate who claimed that local gentry impinged upon *fengshui* when it suited them, and therefore the argument in this case was simply a canard.[39]

No agreements exist today to confirm the final transfer of title of these lands to the railroad, but there is no doubt that Gu was successful. In November 1901, Gu wrote a pronouncement that explained that the metal frame of the railroad bridge crossing the island was completed but was not open to public use. He wrote:

> Be it known that the ground behind the Aozhou Academy is what will be known as the Iron Bridge, and make known that it is for the railroad train track and is not a government passageway, and that it is not allowed for people to arbitrarily walk upon it or obstruct workers or engineers. Furthermore, the bridge does not have wooden planks arranged across it. Supposing that someone lost their footing; that person could be killed.[40]

Sometime later, the bridge was fully operational as confirmed by the map found in the Republican-era gazetteer that clearly shows that the railroad crossed over Ao Island at about the center of the landmass.[41] Currently, the railroad crosses over the island at the same location, leaving the eastern half open for housing while the western half is mostly unkempt and unused.

The case of Ao Island provides a picture of Magistrate Gu trying to find a way to fulfill the demands of the imperial leadership without undermining the goodwill of the local leaders. Gu apparently understood even before the American engineer arrived that the railroad route would travel past the county seat and over Ao Island where bridging the Ping River would be most convenient. He also understood as well as anyone the importance of the academy to the future of scholarly and political pursuits among the elite lineages in the county. To soothe the fears of the local population, Gu turned to some of those educated and well-connected members for support and legitimacy. In the end, when those leaders critiqued his policies, he commanded them to obey and he scolded them and derided them for their parochial mindsets.

Gu's problems did not end there, however. The issue of *fengshui* was important in the burial of the dead, and so the territory used for the laying of rails held particular interest for families whose ancestral cemeteries were in the general line of the route. In fact, selling burial property was so significant that Qing law demanded punishment for the seller that could include banishment or military servitude.[42] Parsons complained that, unlike the cemeteries of the West, graves in China were strewn in a haphazard manner over the countryside making the construction of a straight railroad line nearly impossible.[43] He further explained that in northern China government officials promised to pay extra money to the families for land that contained gravesites as compensation for the removal of the bodies. He went on to complain that while the empire deemed this policy necessary, many Chinese falsified gravesites for the extra money they could receive.[44] In fact, one railroad that linked the iron ore mines in Daye to the Yangzi River was reportedly originally laid out so as to pass by all the local cemeteries, real or otherwise. In part, Zhang Zhidong's assistants made this decision because the owners of the cemeteries were asking for more money than the owners of fields, thus the decision was a cost-cutting one. However, the finished line was so circuitous that trains could not move swiftly enough to make the route workable. Subsequently, the engineers rerouted the line with fewer curves, thus implying that some Chinese agreed to remove gravesites in order to make the line viable. In essence, Zhang's assistants even bought off the ancestral spirits and pushed them out of the way.[45]

These same problems arose when the railroad route was drawn up in Pingxiang County as well. To alleviate the problem in advance a pronouncement was made to all the people of the county to begin preparation for the removal of their ancestors' graves. In January of 1900, a Mr. Xue, the Pingxiang-Liling railroad manager, made an announcement once the engineers finished their initial inspections of the proposed line that announced to the gentry that he planned to "Purchase the Land" and that they should "Open up the Embankments and Remove the Graves." Specifically he proclaimed that all lands along the proposed line were to be measured and seized. All dikes and waterways that provided the valleys of easiest passage for the railroad tracks were taken over, and most lands were to be purchased by eminent domain. Furthermore, he stated that everyone who had graves in the vicinity should begin removing their family plots. To this end, those graves determined to have no family overseers were already removed. However, for those who were overseeing gravesites in the area, in accordance with the empire's regulations, he agreed that for every removed tomb the government would pay 10,000 cash to the families.[46] Subsequently, when W. W. Rich arrived in Pingxiang County to plat out the route, Gu Jiaxiang told him pointedly that they simply had to try to avoid as many burial spots as possible. Rich was apparently agreeable, but he also argued that the route needed to be located along wide and flat lands. He explained to Gu that he would use his expertise to select the route and could not always comply with the wishes of the local population. To collect as much information as possible regarding the local conditions and physical and social hazards, the two men traveled through the western section of the county from the coalmines to the westernmost township around the city of Xiangdong. Along the way, Gu and Rich talked with several lineage heads and others regarding the placing of local gravesites. Based on these responses, they mapped out the line.[47]

Almost as soon as the map was made known to the landholders Gu was informed of a cemetery he did not know existed. After the 1900 Chinese New Year festivals were completed, Gu visited the county prison where one of the prisoners complained that his family's grave was to be overrun by the rails. The prisoner, Li Youxiang, gave the magistrate a letter that Gu assumed protested his innocence. Gu explained that he put the note in his pocket and it was only later that he read the letter and found that instead it was a statement regarding the construction of the railroad. His letter called for a reprieve of the railroad in the interest of local concerns. Li Youxiang explained that the railroad was scheduled to run over his family's cemetery that at the time included his father, his mother, and his father's first wife. This situation was unacceptable, Li argued, and needed to be rethought by Gu and his superiors. Furthermore, Li argued that the plan to allow foreigners to control daily production schemes in the mines and railroad would end in the loss of sovereignty of the empire. To support his argument, Li pointed to a famous case in the Song dynasty when the barbarians from the north were allowed to encroach on the weaker Chinese empire. He also argued that, just like the treaty ports along the ocean coast, Pingxiang County's coalmines and lands were being taken over and would become foreign-held territory, leaving the county and the empire with nothing.[48]

Gu felt the need to respond to this letter openly because it struck a nerve that had been hindering the county's progress. First, he explained that when he and Rich looked over the county for possible routes, they were not aware of the Li family cemetery. When they talked with other local leaders including members of the Li lineage, none of them appeared to know anything about Li Youxiang's ancestors' graves, and so the route was mapped out based on the information available. Gu felt that the Li graves could be opened and moved by another member of the Li family while Li Youxiang was in prison. Second, he reiterated that the decision to construct the mines and railway was made so that China would be stronger and not have to open all its cities to the West. Pingxiang County's industrialization scheme, Gu argued, was under the control of Chinese officials and could not be subverted by foreigners. Only the Chinese leaders had the power to make real decisions regarding the level of development in the county.[49]

More than one year later in November of 1901, Gu wrote his superiors on Li Youxiang's behalf. The magistrate explained that in 1892 Li was accused of banditry with an accomplice. Because the second man successfully fled from the police and there was no one who could be found to formally accuse the men of banditry, it appeared that the prisoner was going to languish in jail without being convicted of a crime. Furthermore, Gu argued that Li proved himself to be a model prisoner and thus his term in the prison seemed unnecessary. Moreover, upon investigation, Gu found that indeed there were many members of the Li lineage buried in the gravesite that needed to be removed. Li's relatives were not among the powerful members of the county, but they were almost certainly part of the powerful Li lineage that included Li Youtang and Li Youru. In fact, Li Youxiang's generational name, "You," signifies that he and his family considered themselves to be part of the lineage and the generation that included the mountain lord Li Youru. Given these circumstances, Gu requested that he be given money to compensate Li Youxiang for the burial lands and that Li be released from prison and allowed to remove his ancestors' graves himself.[50] Gu's superiors granted this request but explained that Li Youxiang had very little time to move the bodies as Rich's time was scheduled to be short. If Li took too much time, he would not be allowed to finish.[51] Thus, even as the issue of gravesite property was a hindrance to Gu's land-purchasing responsibilities, he was able to use his diplomatic skills to satisfy the demands of both those above him and below.

Gu's attempt to fulfill the demands of Sheng Xuanhuai and his railroad administration required more than the designs of the American railroad engineer. Like Gustav Leinung, W. W. Rich and William Barclay Parsons viewed the Chinese landscape as a blank space, and it was their intention to lay rails where they were most suited. For Gu Jiaxiang, on the other hand, the demands placed on the county magistrate required finesse and political savvy. Gentry figures like the Wens had to be brought into the discussion when infringing on the county's most important school and library. Similarly, pushing family gravesites out of the way necessitated a willingness to at least appear to be agreeable to compromise. The final product was to be the reterritorialization of the landscape. But the means by which this was to happen mattered to the community and the government officials that oversaw that transformation. As will be shown later, however, even as the means of sale by the county magistrate followed government practices, Sheng's administrators who controlled the coalmines did not follow these dictates. For them, local politics appear to be much less of a concern and their tactics of purchase were completed accordingly.

Irrevocable Land Sales: Leaving the People Outside Looking In

Zhang Zanchen's Mine Bureau and Gu Jiaxiang's county yamen forced the local property holders—that is, the owners of subsurface or subsoil rights—into coerced transactions. In Pingxiang County, as was true of much of southern China, landlords' property tended to be scattered about the county, with small and large pieces in various shapes and locations. They held subsoil rights to the land that meant that they had the right to the rents garnered by the fields, and they had the rights to sell their property to anyone they wished. These subsurface rights changed hands on a continuous basis as landholders altered their holding portfolios to provide them with short-term capital in one case and long-term investments in another. Conversely, much of the land the government purchased was controlled by a second family that held the surface or topsoil rights. In many cases, peasants simply worked the property and paid rents agreed to with the subsurface holder. In other cases, including many in the region around Pingxiang County, however, peasants attained the surface rights to the property, rights that were as legally significant and understood as the subsurface rights of the landlords. Peasant families often established these rights by paying a sizeable fee to the landlord. Once they paid the deposit, they had the right to use a piece of property essentially forever. This relationship assured the family that the landlord could not remove them for any reason. Even more, the annual fees paid to the landlords by the holders of the topsoil was lower than those who paid rents and did not hold these rights. As with the subsurface rights, topsoil rights were sold back and forth among peasants, though it seems likely that more peasants sought to hold on to their topsoil rights to reduce the burdens of moving between fields from year to year.[52]

Perhaps reducing the impersonal nature of landholding and rental agreements between lord and tenant was the fact that these rents were often established between familial and lineage members. Since landed elites from a particular lineage often lived side-by-side with land-poor members of their own lineage, rental agreements of various levels tended to remain within the familial bond. Even more, economic and agnatic ties overlapped with semi-formal political structures as the state usually required local gentry to perform local protection, taxation, and other services in their communities.[53] In this manner, while the descriptions of the documents discussed later may appear as little more than perfunctory economic transactions, in fact, I argue that there were a myriad of ties and relations between the soil and society. Forced land transaction meant that some social ties, economic strategies, and even family bonds were altered or severed leading to many unforeseen consequences for the community and the state.

Sheng Xuanhuai called upon Zhang and Gu to secure the subsurface rights from the owner of the land rather than the rights to act as tenants of the property. While officials forced these purchases on the property owners, the documents available indicate that the exact terms of the sales were slightly open to negotiation. Government officials, usually unwilling to take no for an answer, did employ strategies that allowed them to take land without causing significant rifts in the society. What was particularly troubling to the local population about these land purchases was that they forced the holder to sell land they did not wish to sell without the opportunity to repurchase the lands as was common for most "conditional" or redeemable land sales in premodern China. Under these contracts, Gu forced the poorer landholders to give up properties they relied upon for their own subsistence. They were not given any opportunities to repurchase the properties even if they had sufficient funds for the transaction. Perhaps even more importantly, elites who were forced to sell their subsurface rights lost not only the land but the right to rents from the past holders of their property's surface rights. This reduced their incomes and in some cases weakened the patron-client social bonds that were so important to lineage power in the county.[54]

Throughout much of the late imperial era most Chinese land sales contracts explained that the transaction was a conditional sale (*dian*), meaning that the seller intended to repurchase the land at a later date. While the earliest Chinese contracts dating to the Zhou era do not contain this term, certainly from the Tang/Song centuries the idea was almost universal that selling one's ancestral property was only to be a short-term plan to acquire money until the person was able to regain sufficient funds to buy the land back.[55] During the Qianlong era of the Qing dynasty (1736–1796), the court announced that all conditional sales could be revoked by the seller after thirty years if the seller could afford to repurchase the lands.[56] In rare cases—primarily when selling gravesite property—the contract was deemed irrevocable under Chinese law (usually written as *juemai*). In either case, during the latter imperial eras, contracts often followed set patterns written in manuals published by officials and others used by local scribes. From these conventions, most contracts followed predetermined outlines and contain similar wording.[57]

However, while the Pingxiang County Mine Bureau contracts maintain these broad standards of language and form, many of them clearly state that they were to be irrevocable sales (in these cases, written as *dumai*). In this case, the imperial government was unwilling to allow for the repurchase of the properties by the current landholders. For example, the body of one contract dated 1899 begins: "This establishes an irrevocable sale of irrigated lands and other properties. The person who drew up the contract, Wang Zhenggui, because he needs the money to assist his parents, decided to sell lands his father had bought."[58] This contract goes on to state very clearly that after the bureau handed over the money for the land, the original buyers could not repurchase the property. Another contract, probably written in 1898, similarly begins: "This contract establishes an irrevocable sale of irrigated fields. Mr. Li and his wife née Zhang, because they have now moved away, wish to sell property located in the area known as the [General Horizontal Alleyway] in [the town of] Anyuan."[59]

These passages raise several issues regarding the agreements made between the local population and the state officials. First, as with nearly all the contracts I have collected, they begin with the same characters stating that the document establishes an irrevocable sale (*dumai*). Given that all of the contracts for the Pingxiang County mine property were dated in the years 1898 to 1900 and were no doubt written by hired literati to assure both parties of their assured rights and responsibilities, it appears that either they were written by the same person and had linguistic quirks unique to that individual or that the use of this phrase was common to this area.

Second, each contract begins by stating that the current owner wishes to sell their property due to some personal reason. While it is impossible to examine the legitimacy of these claims, it was common in all Chinese contracts for the property holder to explain that they were selling their property due to lack of funds. In fact, in the Mine Bureau contracts a statement of that tone was found even in cases where the current owner had several landholdings and showed no signs of economic stress. For example, in one property agreement of one Wen Pinshan, a man who, the documents show, was likely among the wealthiest in the county, the contract begins by stating that he sold his land due to need of money.[60] Other sources add further doubts regarding the acceptance and acquiescence of the local population as the state began purchasing their lands. Gu Jiaxiang's collected letters, for example, include a petition dated 1899 written by a local *xiucai* named Li Shaonan who argued that

> The method of purchasing throughout the province is the same: the [official] who wishes to buy [the commoner's property] selects only the lucrative lands and does not ask [if they want to sell]. . . . The buyer decides on the officially fixed price and commands the landholder to accept the price and draws up a contract. This causes other such people to fail. The property value in *taels* is illusory.[61]

Li's letter here indicates that these contracts might have been little more than bureaucratic fictions and that they were meant to provide official sanction for coerced purchases of substantial portions of the county's resources. In fact, it is likely that state officials used drawn up contracts to legitimize their land grabs in part because under imperial law all privately claimed land was to be held by official deed required to determine legal ownership and tax liability.[62] In this case, government officials forced the local population to write up or pay to have written suitable contracts indicating their willingness to sell and their compliance with all the laws of the empire. These contracts were designed therefore to hide the coercive nature of the deal. At the same time, it is interesting that the officials felt it necessary to make individual contracts with each and every owner of the lands they needed for the construction of the mine and railroad. This suggests that the officials never doubted the status of landholding and its proper place in the social and political realm and that their positions as officials of the court did not supersede that status.

Li Shaonan also points to a second issue, namely, that the state bought up only the best lands and mines. The Mine Bureau purchased the coal deposits that the German engineer determined held the lowest ash content, while the county magistrate was called upon to purchase the lowlands that the American engineer determined were best suited for railroad lines. Given these actions, the approach of the officials appeared to the local population as nothing more than a cynical ploy to take the best the county had to offer and leave the shale and hills for local subsistence.

In fact, the gazetteer describes similar problems experienced by one of its most illustrious figures. The powerful lineage leader Li Youru was an accomplished member of the gentry who held several official positions before he retired from office. As was mentioned in chapter 2, when he returned to Pingxiang County, he purchased an undetermined number of mines and began to extract minerals. However, after Li succeeded in developing the mines, they were purchased by Qing officials and the minerals were sent out of the county. Li's resistance to this purchase indicates that this was a forced sale and that afterward he viewed the Mine Bureau and the government officials who oversaw the agency as exploitative. In a eulogy for Li Youru reproduced in the gazetteer, the author wrote that after the mine bureau bought the lands in and around Anyuan, they

> restrained and sealed up the local mines and created a monopoly so that they inherited the rights to profits of the people's natural resources. They regarded the natural resources [*luan*, literally, "meat"] as forbidden and would not let others touch them. More than several tens of thousands of people were left outside looking in toward the borders of the mine.[63]

In this eulogy the author makes clear that not only did the Mine Bureau officials and the county magistrate take the good lands from the local population—in fact, even the most powerful members of the county—but that once the contracts were signed and filed, the people of Pingxiang County felt abandoned by the court's officials. Attempts by the owners to negotiate for partial access to their former properties were prohibited. To make this point official, Gu Jiaxiang sent out a pronouncement directly threatening punishment for those who tried to use the land. It stated in part:

> Be it known that encroachment on government land is a great offense and a prohibition. The Ping[xiang]-Li[ling] Railroad . . . which is calculated as 35 *li* of cropland, has been properly contracted. . . . In the spring, foreign officials will return to Pingxiang County and they will flatten the

> land. . . . This preparation will be done during the spring in the blink of an eye. The workers of the plows may be ignorant [of the boundaries] and the farmer relies on seeing the local area without constructed borders. It is easy to get confused. I must move from one place to another, and improper occupations of the land will cause delays of our important work. I will appoint gentry to investigate and oversee and discuss each of the plats. The gentry *bao* leader will also hand down orders to each family head and lineage. A placard in the neighborhoods will instruct each farmer and each person to act in accordance [with this pronouncement].[64]

And so gentry leaders and village heads, landlords and mountain lords, could only watch with disgust as their mines and paddy fields were bought up, cordoned off, reorganized, and made more lucrative without them.

At the same time, the economic and social pain caused by these transactions may have been partially alleviated by the nature of landholding and tenancy relations in China. As was stated earlier, most landholders owned many pieces of property located throughout their home townships or localities. Unlike European manors that tended to be centered on a single large field, Chinese land strategies meant that even if landholders were forced to give up a particularly valuable tract, they would not be left landless. They almost certainly had many other fields in locations away from the railroad route that continued to gain rent. Furthermore, for landholders who were forced to sell their properties but wished to continue to live as landholding gentry, it was easy and common for elites to purchase other fields and continue to live off the rents of tenants. Moreover, tenant farmers who lost their topsoil rights dues to government purchase could simply establish new tenancy relations with other landholders or with the same landholder working different fields. Since both surface and subsurface rights were fluid in late imperial China, eminent domain takings by the government did not necessarily lead to severe hardship for either lords or peasants.[65]

Regardless of the fact that the documents fallaciously stated that the seller was a willing participant, it seems likely that the financial terms written in the contracts tell us the actual price the Mine Bureau officials paid for each of the properties. The prices the state officials were willing and able to pay—and the prices the local landholders were willing to accept—for the construction of this mine and railroad gets at the issue of the state's perceived obligations toward its people regarding acts of eminent domain and the length to which the state would go to succeed. While the original Qing code was not clear regarding policies pertaining to official purchase of commoner property, it was revised more than once to clarify this gap.[66] By the twentieth century, the court stated in the post-Boxer reforms that the price of land purchased by court officials for the purposes of laying railroad track was to be fixed to avoid speculation. Specifically, the reform legislation stated that in order to hinder local attempts at extortion or inflation, the local officials could not raise prices.[67] It is likely that the court was aware of the variation of land values in different parts of the realm and so no fixed price was given. However, the assumption was that a fair price could be found based on the quality and quantity of the lands being purchased and that a fair price should be paid to each landholder. In Pingxiang County, the price that the local landholders hoped to receive and the amount of money the state believed was proper led to several conflicts between buyers and sellers, mostly eventually resolved through the exertion of state power.

There is no doubt that the land purchases in the county were extensive. With the route estimated at about 15 miles, County Magistrate Gu Jiaxiang's Land Purchase Bureau purchased a piece of contiguous property from the people of the county that was sufficient for the rails.[68] This was almost certainly more than any gentry family in the county held at the time. Furthermore, because the railroad was to travel mostly along expansive flat terrain and near large markets and cities, the properties under consideration were among the county's most lucrative and expensive. Gu realized that all the land would cost a great deal, yet he also knew that to spend heavily on the land for the route could make the construction of the branch line prohibitively expensive. And so, his purchases had to fit within some form of a budget, though it is not clear what his bottom line was.[69] Operations of this size were so complicated and the revenue and expenditure rates so convoluted that few bureaucrats like Gu could know what their expenses needed to be to make a profit. Furthermore, in the case of the "Official-Supervised Merchant-Managed" Hanyang Ironworks and its subsidiaries, the profit motive may not have been necessary at all given that investment capital was coming from taxes as well as foreign loans. In any case, Gu did write that cost was a concern, though he did not provide the details to indicate what he felt he could reasonably spend.

The budgets for County Magistrate Gu's Land Purchase Bureau and Zhang Zanchen's Mine Bureau were also affected in part by the market price of the land. Both men were aware that by court command, land purchases had to fit broadly within a locally accepted price. For Gu, it was imperative that he not be too heavy handed for fear that he would cause unrest in the county under his charge. To determine the proper price the government should pay for the local lands, Gu and Zhang did an extensive investigation and apparently acted accordingly. Furthermore, Gu and Zhang sent men into the countryside to determine the value of the local property. Based on these investigations Gu determined that the standard *mu* of land was about the same as 20 hands (*ba*).[70] These investigators also determined that each *mu* produced two crops of rice or grain and thus averaged 8 or 9 piculs of rice per year. Gu further calculated that for many peasants half of the crop was given over to the landlord as rent, leaving about 4 piculs of rice for the renter on which to survive.[71] Echoing these figures, one local landholder stated that his best properties brought in annual rents of about 4 piculs per *mu*.[72] Taking his numbers, rough as they were, Gu attempted to determine the total value of the lands he hoped to purchase. He calculated that the value of land that produced 1 picul of rice was about 14,000 cash. Thus, Gu argued that the land was worth about 60,000 cash per *mu*. However, Gu also noted that in a similar circumstance around the Daye Iron Ore Mines in Hubei Province, the magistrates paid local landowners about 26,000 cash per *mu* for their land. He believed that his much higher calculations, rather than the 26,000 cash offered in the iron ore mines of Daye County in Hubei Province, were the correct figures he would have to follow.[73] Given that the official conversion rate was about 1,000 cash to 1 foreign *yuan* (or 1 Mexican silver dollar) this works out to about 60 *yuan* per *mu*.[74] These calculations fall nicely within the broad outlines of Jerome Ch'en's assumptions that in the early 1930s in the highlands of western Hunan and Hubei and eastern Sichuan, "intermediate"-quality land paid about 35.5 *yuan* per *mu* and the "highest"-grade land paid as much as 120 *yuan* per *mu*.[75] This suggests that Gu was trying to pay the proper market price for the county's land while other officials elsewhere may have been paying less.[76]

Even though Gu was convinced that the land was worth more than the Daye officials spent, he was hampered by the fact that he did not have sufficient funds to pay for all the land in the few months he had before W. W. Rich was to return. Gu apparently planned to pay the higher figure of 60,000 cash per *mu*, a price that would seem prohibitive for a county government or even his Land Purchasing Bureau.[77] He suggested that he pay the peasants half of the price for their land at the time the contract was agreed to and would pay the rest on a fixed date after harvest.[78] This payment schedule coincides with the land tax schedule used in the late Qing dynasty that demanded that half of all taxes be collected by late spring and the other half after harvest.[79] He also surmised that, given the local form of land measurements in Pingxiang County, there would be fewer problems if he used local methods for land purchases. In fact, Gu argued, given the land roles as they were throughout the county, he would not know how to take these measurements. Gu's superiors, however, rejected his proposal and informed him that all purchases of property were to be in *mu* of land and not in local methods of measurement.[80]

Under these constraints, Gu set out to buy the lands he needed with contracts that were agreeable to both the owners and his superiors attempting to pay the prices he felt were proper. However, even Gu's higher asking price proved to be problematic and led to conflicts with the local landholders. Having received Gu's offer, a *xiucai* named Li Shaonan wrote a pointed letter in which he argued that the purchase price Gu was suggesting was simply too low, given the value of the land as a source of income. He apparently explained that because his family's land was valuable, he received 4 piculs in rent every year, making for a calculated value of 50,000 to 60,000 cash per *mu*. Thus, Gu's purchase price offer of about 60,000 cash per *mu* was not acceptable. Under these circumstances, the landowner argued, the land's purchase value was not much more than its yearly rent intake.[81] As with so many of the conflicts the documents describe, it is not known if Gu changed his offer in any substantial way. However, the body of information suggests that he likely did not. Gu was ordered to secure all needed lands and all evidence suggests that he fulfilled his obligations. Even though no one seemed to know what the budget would allow, Gu's purchase price was likely based on a negotiated settlement between the market prices and what he as magistrate could pay.

What is certain is that County Magistrate Gu Jiaxiang was given the responsibility to purchase a swath of land that stretched from the center of his county some 30 miles to the western border that was to facilitate a new transportation system suited to exporting millions of tons of the county's raw materials that had been a key component of the local economy and material life. Even though he was deputed for a second term of office—in fact, the county leaders requested him specifically—he was called upon by the state to act as a collaborator in the forced reorganization of the county's property and economic topography. Several documents make it clear that even as Gu extracted lands from the elites as he was commanded by court officials, these actions also undermined his credibility with the county's lineage leaders who assisted him in many of his local duties.[82]

To alleviate some of the conflicts with the people of the county, Gu sought out the best strategies to both placate the local population while performing his duties. In one petition to his superiors, for example, Gu explained that construction required to support the railroad lines in Pingxiang County, such as embankments and irrigation ditches, was more harmful to local paddy fields than was the case in other, flatter parts of China. He wrote in part:

> Recently, we carried into operation [a railroad track] much like what was decided in Hubei and Jiangsu [provinces]. But in the north, the conditions versus the south are not the same. The bank of a river and the bank of a lake's area within the interior lands are not the same. In the north, a piece of level land along the government postal routes was originally expansive. With the rails it was possible to follow these posts in order to make it work. Also, every now and then, we dug a ditch or the postal routes needed a small bridge. The outlines of the tracks were all evident. In the southern lines, the mountains always have crooked trails. The land slants between the fields and the raised paths act as dikes with river beds surrounding and enclosing them. The narrowest [pathways] are perhaps a foot [*chi*] and the widest are several feet. The railroad is most valuable to us where it is wide and flat and upright. Is it possible for us to make this happen? Ultimately, I had to myself develop the field embankments and [to this end] I had to repair the walls completely. Consequently, I encroached upon the people's fields. I blocked them up completely and changed the boundaries. The banks of rivers and the banks of lakes are very different. Here the waterways often overflow and the people fear they will suffer. They guard against [this problem] as if there were bandits in their lands. Around Gaokeng, [on the other hand], the water is always insufficient. Since they can gain from what we have, the smallest drop of water is precious and beautiful. I have heard that the railroad [engineers] greatly fear flooding. And therefore, [officials from Hubei] discussed raising the embankment and fixing the shore line. Then we must change the rules regarding the railroad and the people's fields, and we can rely on the railroad to defend our mutual desires and ward off calamities. . . . We can establish waterwheels for irrigation, use wheels for river flow to push the water to turn around and then make the river water flow into [the Gaokeng area], and the stream will go into the fields.[83]

In this petition, Gu is trying to be conscious of the needs of the people of the county. He hoped that his empathy, together with some clever planning, might allow him to actually make some people's lives better. Even if he did have to take land away from the people, his plans could solve age-old problems of the county's dry regions to the east and excessive flooding in the west and therefore act as salve on the wound. However, in the end, whether he or anyone else in the county liked it or not, the land purchases were a court-mandated priory that he simply had to fulfill.

While the story of Gu Jiaxiang and his Land Purchasing Bureau suggests attempts at shrewd political negotiations partially concealing a heavy hand, the Mine Bureau contracts indicate that the mine officials were not as interested in local concerns. In fact, the local population apparently considered the purchases of Zhang Zanchen to be particularly brutal and unfair to the landholders.[84] In one irrevocable sale, the sellers explained that the property was 155 *ba* and that the rent charged on the land was 115 piculs or twice Gu's highest calculation. For this land the government agreed to pay 310 *yuan* in foreign currency.[85] This calculates to only 40 *yuan* per *mu*, a figure lower than Gu's 60 *yuan*. Another contract for irrevocable sale gave the property as 150 hands—or 7.5 *mu*—with a price of 110 taels of silver. This very low price fits into calculations that the poorest lands in the region were purchased for as little as 6.8 *yuan* per *mu*.[86]

Unfortunately, the rest of the contracts I examined did not provide sufficient information to determine a breakdown of the price of the property with this level of certainty. As was stated earlier, one contract for three mine shafts on the Zijia Mountain was signed by a member of the Peng lineage, for which he was to receive 45 *yuan* in foreign currency.[87] Similar to other contracts, this is probably less money than Gu's payment scheme allowed as the total property according to Gu's figures would have to be less than 1 *mu* of land, a size that is possible but unlikely for three mines.[88]

Also, as was mentioned earlier, when the government wished to purchase gravesites, the court decided that a compensatory fee should be paid to the landholder for the labor to remove the bodies.[89] In one case involving gravesite property, one of the Mine Bureau contracts depicts a sale by one Wen Xianli, who sold his family's graves located in a bamboo grove near the coalmining site of Gaokeng. For this sale, Wen received 140 taels or just over 100 *yuan*. The contract does not tell us how large the property was, so it is unclear what if any extra compensation Wen may have received for moving his ancestors from the property.[90]

Anyone who has tried to determine with any certainty the price of goods and services using the many forms of currency being used in late imperial China knows that these calculations are only suggestive. Also, I must assume that accuracy of spatial measurements regarding field sizes were more indicative than real. However, I would argue that these figures tell us several things. First, it does appear that the state was paying a price that was near the true value of the land. Even if we take the lower figure of 10 *yuan* per *mu*, this is close to prices paid out regionally. Moreover, the contract that paid about 20 *yuan* per *mu* may provide us with a benchmark for the fixed price called for by the court. At the same time, the landholders who complained to Gu Jiaxiang may be correct when they argued that the price was simply too low for the lands they held. Because Pingxiang County was located near the eastern Hunan provincial rice marketing city of Xiangtan and several other expanding markets—not the least of which was soon to be Pingxiang City itself—one could argue that the value of land was higher than in the more remote areas of central China where Jerome Ch'en completed his calculations. Indeed, the letter to Gu complaining of the price argued that the quality of his land was particularly high and therefore he received much higher rents than others.

This chapter has shown that as government officials hoped to reorient the county, local inhabitants pushed the state to make concessions along the way. The Wen lineage leaders, in particular, engaged in a series of agreements with the Mine Bureau in the hopes of buying time needed to make improvements to their firm. However, the lack of capital and technological know-how kept them from pleasing Sheng Xuanhuai and the managers of the Hanyang Ironworks. The Wens were not the only gentry members who tried to push the state in various directions. As the railroad was platted out, local literati opposed plans to use the Ao Island for the bridge across the river. This property represented for them a tangible edifice of their status and power, and the railroad indicated that their needs were simply not significant given the demands of the local coalmines. Moreover, as landlords fought over the price of land, they were not only negotiating a price but also determining the relative value of landed elites versus the coming railroad. They had to accept that their role as the holders of surface rights on paddy fields was not as useful to the empire as the court's ownership of railroads and coalmines. And along with this loss of property came the diminishment of their status as the holders of the community's land and labor, and this harmed not only the local economy but also the social fabric that was important to the county's socioeconomic conditions. In fact, in the case of Li Youxiang, the man who wanted to retrieve his buried ancestors, the forced purchase of grave land signified that the state determined that modernization was more important than the sanctity of the family.

Finally, it should be stated that the reterritorialization that was taking place in China sought to change the social and economic structures of the county. The collection of coalmines that dotted the mountains and provided subsistence to the people were gobbled up and centralized in the hands of outsiders. Moreover, centuries-old markets and transportation routes that facilitated the distribution of coal and other goods to the local inhabitants were done away with and replaced by a single transportation line that was created not to provide for the local community but to service factories whose purpose was beyond the scope of the county's desires. In essence, these land purchases and industrial designs cast the local population as expendable if not a nuisance to the state. And as the technology was integrated more completely into the greater coalmining scheme, the ways of life among the peasantry and landlords became even more of an anachronism and less of a significant part of daily life in the county.

5

Mechanization of the Coalmines

Tearing Down and Building Up

Mechanization of something as complex as a full-scale coalmine in the nineteenth century required many technological components that complemented one another. But it also necessitated political and social adjustments that conflicted with the needs of various sectors of the community, Chinese and foreign, workers and elite, official and commoner. In the previous chapter, I showed that beginning in the late nineteenth century the state grabbed control of the land that had been in the hands of lineage leaders for generations. With these properties under the ownership of the state, landlords lost their rights to the rents and similarly miners could no longer turn to their lineage leaders for their safety net. Chinese officials and foreign engineers became the new landlords. These new bosses hoped to impose a new technological order on the land that would greatly increase coal extraction from the mines. To this end, the German engineer Gustav Leinung and his associates had to facilitate technological transfer, a term that implies a simple introduction of machinery to another country but that often entails much more. In fact, by the late nineteenth century—the so-called second industrial revolution—technology needed to assemble an entire industrial scheme requiring at least a rudimentary knowledge of modern math and physics, electronics, steam power, and the working of the internal combustion engines, as well as the ability to impose the industrial landscape into the receiving country's indigenous culture and sociopolitical world.[1]

The politics of industrialization necessitated the integration of at least three different actors into a mixed pot, each of whom brought their own complementary and conflicting needs to the project and each of whom could therefore hinder or even defeat the project.[2] First, at the most macro-level, the Western powers presented their vision to other parts of the world in the guise of engineers sent to dismantle the local community's premodern methods of production and implement new ones even at the expense of local understanding of the environment.[3] Second, governments that purchased or accepted Western technology into their otherwise pre-mechanized world hoped to advance their material standing often at a cost extracted from the state's coffers and their natural resources. In times of empire-wide crisis, states often attempted large reforms designed to address the perceived source of the problem. National initiatives focused on the macro-economic and military needs of the state.[4] And, finally, once foreign powers and governments made the decision to begin a project, workers and consumers in the locality were incorporated into the scheme as well.[5] To this end, patronages and premodern economic and social systems were subsumed into a new order and direct kinship bonds were severed. Workers maintained some contact with their countryside brethren through these institutions while the gentry and peasant community alike maintained tangential connections to their kin.[6]

The collective desires of the fragmented community as well as the divergent demands of a growing working class pushed against the goals of the state and the foreign engineers. Similarly, the foreign engineers who needed a skilled labor force and capital-rich investors were forced to alter their strategies and accept less than a complete effort as state and local monies failed to keep up with the demand for more expansive technology. Since industrialization required the integration of these sectors into a complementary structure, the fissures within the system altered the project and its chance of success. It is this conflict that I address in this chapter.

Constructing Anyuan: The "Flourishing and Fast-Paced" Town

After the Mine Bureau and the county magistrate purchased the lands, the German engineer Gustav Leinung completed the broad outlines of the scheme. He centered the chief activities within the town of Anyuan near the mining sites of the Wen lineage personal holdings, the collective holdings of the Guangtaifu Lineage Trust, and countless other smaller mines owned or operated by Pingxiang County landlords and commoners alike. This town grew to be an amalgamation of Chinese and Western culture as foreign engineers negotiated with local workers over such issues as the technology and mining practices, labor training and expertise, social connections and subsistence requirements.[7]

Beginning around 1900, Zhang Zanchen and Leinung essentially built the new city of Anyuan virtually from the ground up, turning what had been a collection of working and abandoned mines into one of the most prosperous and modern cites in all of Jiangxi Province.[8] As the 1935 gazetteer suggested:

> Anyuan is a special market town 10 *li* south of [Pingxiang] City. The adjoining village was completed after the mining works developed and the plant and manufacturing buildings were expanded. The wall and city gates [enclose] workers and merchants who are crowded [together], and added to that are the railroad, communications, and trade that are flourishing and fast-paced.[9]

This entry shows that Leinung and the Chinese managers constructed the city almost from the ground up and with a great deal of precision and forethought. Unlike most towns, county seats, and formal cities in China that opened their gates to the peasant landscapes and periodic central place-based merchant trading routes, Anyuan was designed to be a node of production for a bustling imperial-wide economy.[10] It recreated a Western company town centered on the large horizontal and vertical openings dug into the mountains' belly, with a collection of buildings suited to extract, purify, and distribute coal for industrial purposes. Spatially, it held all the buildings and machinery needed to provide for the extraction and refinement process. Walls separated the entire mine complex from the rest of the countryside, much like a county seat. Temporally, the county seats followed the seasons and planned meetings and political functions around the calendar, but in the mining city a whistle went off every twelve hours to indicate the beginning of a new shift. Unlike the bureaucratic center that contained both the administrative yamen and the gentry districts and included a high percentage of silk-robed gentry members riding on sedan chairs, workers in dungarees with blackened faces as well as more smartly suited merchant managers and foreign bosses populated the factory town of Anyuan.

In the decade after 1896, the coalmining scheme in Pingxiang County gradually developed along Western lines with new modern equipment. From the earliest days of the project, two methods of mining continued throughout the period: the continuation of Chinese vertical shafts and the development of modern horizontal shafts. In order to assure the state of at least some output prior to mechanization, Leinung and Zhang Zanchen ordered the best of the older vertical shafts to continue to function, and workers maintained them as they had before. The two most important mines of this type were named Bafangjin and Liufangjin. Bafangjin went about 500 feet deep, reaching the fourth layer of coal, while Liufangjin went 300 feet deep, extracting coal from only the third layer.[11] Brick walls along the main shafts and, in some places, steel frames protected the pits from collapse. Ventilation bellows supplied air deep in the pit, and water pumps kept the pits from flooding. Miners followed the seams of coal for over 100 meters from the well as they undulated through the subsoils. The extracted coal was then hauled up initially by horse-drawn carts and in later years by Western-made winches. These two shafts, utilizing modern and premodern labor and technological systems, produced 600 to 800 tons of coal per day.[12]

At the same time, Leinung greatly expanded the use of Western technology in newly constructed horizontal shafts, or adits. These tunnels connected several of the old Chinese-constructed shafts together and employed electric trains for transporting tons of mineral out of the mines. Almost as soon as Leinung took control of the mines, he began writing Sheng Xuanhuai, requesting the rights and tools to excavate a series of horizontal shafts linking several of the purest deposits together. In a letter dated October 10, 1898, Leinung wrote a memo to Zhang Zhanchen, Sheng's Chinese appointee in Anyuan, requesting the construction of such a series of mine adits. He explained that the longest mine adit would measure about 2,500 *chi*,[13] and that a second shorter adit would likely stretch some 1,000 *chi* and be located at a higher elevation than the longer one. Both would open at the base of the mountain range and travel roughly southeast toward the most lucrative deposits, including the Zijia and Tianzi mountains.[14] Leinung explained that Sheng Xuanhuai asked him to find the deposits lowest in phosphorous. Leinung concluded that the Tianzi deposits were the best quality, making them the most important.

While he emphasized the necessity of horizontal adits, he also listed two problems that had to be solved in order to complete the 200-*chi* leg. First, he mentioned that, even as constructing adits allowed for greater use of modern equipment, they might also cause political friction because the new tunnels joined together the old mines that had been worked by different families and mountain lords. These shafts were the subject of feuds and competitions among miners for generations. "I fear that the local people will obstruct this project," he wrote, "and we can avoid this project. But wouldn't that just hinder us that much more?"[15] Second, he explained that he needed more foreign engineers and Western equipment to do the jobs that the local laborers could not. The machines required included jack hammers, electric air ventilation systems, and modern scrubbers and coking ovens that could only be operated by skilled laborers. And so Leinung envisioned modernizing mining in Pingxiang County by replacing the miners skilled at premodern methods with miners skilled in modern technology. These two changes promised to greatly increase the output of the mines by dramatically altering the production scheme.[16]

Given these two caveats, the adits were indeed constructed much as Leinung requested. In fact, he excavated not two but three tunnels linking the best deposits. The openings came together at a focal point where he planned to construct the coking ovens and scrubbers. The workers and engineers called the new adits the "General Horizontal Alleyway," which was the longest adit at 3,000 meters; the "Eastern Horizontal Alleyway"; and the much shorter "Western Horizontal Alleyway." Each contained alleyways and pits that employed the room-and-pillar as well as the longwall method of mine excavation.[17] Inside the rooms beneath the surface, men used a combination of picks and hoes as well as modern equipment to break the mineral loose from the mountain rock. In the General Horizontal Alleyway and the Eastern Horizontal Alleyway an electric winch hoisted the coal into a cart drawn by China's first electrical train. These one-hundred-hauling-car German-built "Dynamo" trains operated by 1906 with motors that exerted from 10 to 50 hp. The trains traveled the full length of the adit along brick walls that were lighted by electric lamps 15 meters apart. Another electrical cord followed the alleyway that supplied the train with electrical power.[18] Even after the electric trains were operational it appears that digging continued to extend the alleyways further into the mountains until the adits were completed by 1908.[19] With this effort, Leinung successfully integrated modern technology into the county's coalfields and he replaced the complex property delineations of the past with a unified mining scheme.

However, even as Leinung used more efficient methods of extraction, the Hanyang Ironworks managers still deemed the quality of the coal insufficient for smelting purposes. Chinese miners apparently extracted good-quality coal out of the mines but then added heavy shale to increase their pay. Similarly, mine workers failed to adequately purify the mineral, employing the Chinese coking process. On February 3, 1899, Sheng Xuanhuai sent a pointed letter to Zhang Zanchen explaining that Eugene Ruppert, the foreign manager of the Hanyang Ironworks, found the quality of the coke to be inadequate for their needs. He directed Zhang to force the workers to be more diligent in their efforts. "If the workers still cannot separate the coal properly," Sheng stated, "the managers should punish them."[20] Sheng argued that the coal was of poor quality and of insufficient quantity, demanding 1,300 tons of high-quality coke every month.

In March of 1899, Leinung and Sheng Xuanhuai continued this conversation in Shanghai. The German engineer explained that the mines needed more equipment to increase output and solve the purification problem. Some of the machines he required were purchased earlier for the Hanyang Ironworks or were constructed at the Ma'anshan Coalmines located in southern Hubei Province but run by the same engineers and officials who operated the mines in Daye more than 100 miles away. Since foreign engineers determined that some of those tools were no longer needed in the Hanyang factory and that the mineral in the Ma'anshan Coalmines was insufficient, he argued the firm could ship the tools to the Pingxiang County mines. Leinung also told Sheng that there were several important pieces of machinery that had to be purchased. Sheng told Leinung to draw up an official list written in English and Chinese, and he would take care of it.

Almost immediately, upon returning to Pingxiang County, the German engineer wrote up a lengthy list of the machinery required for further development.[21] One of the most important pieces of technology Leinung requested and then received was a coal scrubber used to separate the coal from the shale. With this device, the extracted mineral could be taken out of the train cars and placed on a conveyer belt just outside the adit, and from there sent to buildings about six stories tall. The coal was sieved through a series of screens and then sent to the scrubber, which is a metal vat of water that swirls around in a whirlpool. As the water briskly swirled around, the lighter coal rises to the top of the machine, while the heavier shale and dirt fall to the bottom where they are extracted through a hole underneath.[22]

The scrubber separated coal from the shale and other waste products and then ground the coal into small pieces. The coal was then loaded in electric-powered trams and sent to the coking ovens nearby. As noted earlier, the Guangtaifu Lineage Trust operated Western coking ovens even before Leinung arrived. After the Wen lineage bought the first oven, they reportedly purchased seven more Western ovens that may have worked but certainly did not fulfill the Hanyang Ironwork's fuel demands. Zhang Zanchen informed Sheng that he could send as much as 1,500 tons of coke every month if he could get more equipment for the job. He explained that he and Leinung needed to bring in another Western coking oven as soon as possible. To this end, Leinung immediately left Pingxiang County to see if he could get another coking oven either at the Hanyang Ironworks or the Ma'anshan Coalmines.[23] Sheng argued that they should use the coking oven in Ma'anshan, while Leinung felt the one in the Hanyang Ironworks was less necessary in the immediate months.[24] These letters do not make clear which plan they put into action, but they do show that Sheng Xuanhuai quickly sent a coking oven to the coalmines to increase the output in the county. Also, Leinung added even more coking ovens to his list of requested supplies. By the mid-1910s, three hundred German-made electric coking ovens formed a wall, burning night and day. Alongside those stood an area estimated by one Western observer to be as much as five acres in size that held premodern Chinese ovens. The trenches that were designed to perform the same function as the electric ones remained in operation well into the 1910s and supplemented the Western methods.[25]

Leinung's list of required purchases included several more devices considered essential for running a modern coalmine at that time. Among the items were a large winch and a smaller, mobile, winch, as well as cranes to move heavy minerals out of the deep shafts and at least two pumps capable of extracting up to 1,000 liters of run-off per minute. He also asked for a power jackhammer that could break through the mountain rock. Since he needed at least one brick building for a scrubber, and brick walls for constructing the adit, he requested a brick-producing plant that could produce 2 tons of brick per day.[26]

Once the equipment arrived, Leinung established a factory-like assembly line that moved the coal from one step to another in a tightly centered facility, keeping transportation costs to a minimum and assuring ease of operational control. To the uninitiated, the space near the General Horizontal Alleyway entrance looked like a Rube Goldberg system of trams, conveyer belts, scrubbers, and kilns that purified and coked the coal and then finally sent it on its way. To contain the tools and motors, Leinung erected sheds and buildings. Near the opening of the adits he constructed engineering rooms that housed generators and repair shops as well as warehouses for timber and wooden frames. By the 1910s, Leinung even purchased equipment that constructed and repaired the mining tools and even the trains that arrived at the train station.[27]

Many nondescript white buildings housed any number of pieces of equipment, materials, or offices for mining services, and several boilers and chimneys dotted the factory and the coal town landscape. At the opening into the General Horizontal Alleyway, not far from the train station, stood one of the tallest of the chimneys. Visitors to the city describing their first glimpse always mentioned first and foremost the smokestacks spewing soot and ash as well as the loud booms emanating from the furnaces. The Chinese teacher Ci Fei, for example, described stepping off the train on a visit to the mines. "When I arrived on a very hot day," he wrote, "I walked beside the boiler's cylinders and the blazing hot flames and steam that oppressed the people there. It reminded me of the novel *Journey to the West* with the description of flame-throwing volcanoes."[28] He wrote that even as they were somewhat apparent in the daytime they lit up the night sky. In another instance, one young boy who arrived with his mother and sisters to beg for food and find work told of seeing Anyuan for the first time. When they walked past the shops and cafeterias on the narrow streets with *baozi* and sesame cakes, they approached one of the large smokestacks "that truly meet the sky." At the base of the chimney "were several workers whose heads were covered in large beads of sweat, and right in front was a saucepan oven's high flames where the men shoveled the coal." The sight of modern economic development was powerful to the impoverished peasant family and led them to believe that they could find the financial help they needed.[29]

Outside of the immediate lands of Anyuan, the court granted Zhang Zhanchen and Lu Hongchang the rights to control some of the smaller coalmines. These court officials organized the shafts located just outside the company-owned properties into partnerships. They commanded the miners there, in turn, to sell their minerals to the firm at predetermined prices. Under this agreement, as late as 1935, more than seventy small mines, each employing from ten to several hundred workers, each sold several tons of coal daily to the Anyuan mines.[30] This indicates that even within these more peripheral areas, the Chinese government controlled these production schedules.[31] Some mines that were too far away even for Zhang and Lu's interests also continued to function and provide coal to the local markets. In particular, the mines of Anle Township in far northern Pingxiang County contained coal that was not as high in quality but was nonetheless considered suitable for the mechanization process by Lu Hongchang.[32] Because the mines were far from Anyuan and were separated from the modern town by difficult terrain, Leinung decided to leave these mines out of his modernization scheme. Even so, many small and locally worked mines in Pingxiang County continued to function well into the first decades of the twentieth century and provided both fuel for local consumers and seasonal jobs for some of the county's peasant men.

The Railroad: Delivering Kerosene and Breaking Rice Bowls

Delivering thousands of tons of mineral from Anyuan's plant gate to the Hanyang Ironworks created problems that could not be solved using premodern methods. The transportation costs and acts of political corruption curtailed any hopes of employing the overland and river-based routes previously linking Pingxiang County with the important cities of central China for centuries. As shown in chapter 3, complaints about Guangtaifu-hired boat haulers adding heavy shale to their loads to increase their pay and the corruption of *lijin* offices in Hunan Province convinced the Hanyang Ironworks officials that constructing a railroad would reduce the time needed for transportation and centralize the control required to provide high-grade mineral needed for the smelting process.[33] Therefore, Zhang Zhidong and Sheng Xuanhuai decided as early as the mid-1890s that building a modern railroad and communications system from the coalmines to the ironworks factories was the only adequate response to their transportation needs. In 1898, Zhang completed a deal with the American-China Development Company to construct the line from Pingxiang to Zhuzhou, a small market town along the Xiang River in Hunan Province. Even though the contract called for the Western engineers to complete their task in three years, both parties agreed to a new contract that extended the construction period. Working under this new agreement, by 1903 the American engineers completed the railroad to Liling City at the site of a newly established railroad administration building. By 1905, the entire sixty-five-mile-long route to Zhuzhou, an industrial town along the Xiang River, was all but finished.[34] From the railroad terminus, small boats shipped the coked coal down the river. Once the coal reached the city of Xiangtan, where the river's depths permitted heavier traffic, larger boats delivered the coal to the Wuhan Cities.[35] Once the ships arrived there they had to unload across the river from the ironworks because of a lack of port facilities in Hanyang.[36] With the work not fully completed by 1905 however, the American-China Development Company failed to accomplish the project as they had agreed. Because of this, Sheng Xuanhuai apparently took over the enterprise himself and completed the project for less money than the Americans with nearly equal quality using his Western-trained engineers and managers.[37] Eventually, by the 1920s, the Guangzhou-Wuhan trunk line was largely completed and a reported six trains were daily able to deliver the coke directly from the mines in Anyuan to the mines and factories in Daye.[38] It should be noted, however, that the cost of shipment continued to be a serious problem for the ironworks.[39]

The railroad did more than send coke out of Pingxiang County. Railways supplied the city with many goods needed for production. The transportation of fir trees from Liling County supplied the extensive need for lumber required to support the tunnels and shafts. Chinese workers exploited significant forest supplies on the county's westward-facing hillsides for sale and consumption in the Xiang River basin in Hunan Province.[40] Liling County merchants also sent other goods up to the markets in Pingxiang County, including red tea, ramie, porcelain, Tung oil, and paper.[41]

In the same way that the railway supplied industrial products for the mines, it also brought in personal consumption items from throughout the empire and even other parts of the world.[42] Some of these goods reached Pingxiang County where they influenced local consumption. The young scholar and future Communist Party leader Zhang Guotao wrote that when he entered public school in Pingxiang City, the marketplace was filled with goods he was unable to purchase in his home township.

> Modern transportation and mining brought with them many social changes, the most distinctive of which was the transformation of a tiny shop selling foreign goods into [Pingxiang's] largest store. Foreign cloth, foreign oil [kerosene], manufactured metal items, and other foreign products gradually spread over the countryside, posing a direct threat to the handicraft industries.
>
> . . . Foreign cloth, kerosene, and foreign lamps, moreover were viewed as contemptible fads. Grandfather was outraged when I returned home one day dressed in a gown of foreign cloth; and the lamps and other foreign goods that my father brought with him from Shanghai deeply offended the eyes of our elders.[43]

This description provides a vivid picture of the changes wrought by the coming of the railroads and indicates that it altered the daily lives and consumption habits of the people along the route.

Similarly, train stations were erected for passengers and cargo. There was a stop at the market area near the Lesser Western Gate at Pingxiang City as well as stations in Liling City and Zhuzhou as the line descended toward the Xiang River of Hunan Province.[44] At each stop along the way Chinese station managers hired workers to oversee the incoming and outgoing of passengers and cargo as well as manage the workers who performed mechanical duties required of railroad maintenance. One peasant-turned-railroad-worker, named Wang Zheng, for instance, explained that

> . . . I went to the stationmaster's office at the Canton-Hankow Railway and secured work as a servant. . . . When I was thirteen or fourteen I was promoted to the job of switching rails at seven dollars a month. Soon afterward I became an oil feeder and examined, repaired, and oiled the locomotives. For this I received fifteen dollars. Next I was made locomotive firemen, at seventeen dollars and fifty cents a month.[45]

It is clear from this description that the expansion of Western technology spread into the cities and small towns as poor laborers were hired to perform tasks based on this new machinery. In addition, the worker's recollections of payment and skills illustrates the emergence of a hierarchy of jobs and positions within the railroad system. Unlike family-based boat-hauling firms where the owner and the operator were either in close proximity or, indeed, one and the same person, the railroad required Zhang Zhidong and his associates to trust the actions and dedication of Chinese managers, overseers, and day laborers in many locations along hundreds of miles of track. Men severed their patronages and left the villages behind them as they moved to the towns and cities and developed technological skills so they could live as mechanics and industrial laborers. Thus, the founding of the coalmines in Anyuan and their connecting ironworks and iron ore mines sparked the expansion of social change throughout the region.

Conversely, as the railroad line provided new jobs, skills, and even careers, it also pushed old occupations aside, leaving some families and workers in precarious circumstances. Boat hauling and overland foot traffic occupations that fed some families for generations were eliminated by this cheaper means of transportation. The railroad also altered the costs of transportation along its path, making some firms located near the lines more economically beneficial than other firms that had the ill fortune of being located along old river routes or footpaths.[46] In his memoirs, Zhang Guotao writes that the coming of the railroads sparked unease and facilitated dislocation from many people in the region. As a young man, he learned from one of his teachers that even if this modernization would benefit the country in the long run, there were other negative consequences: ". . . the railroad had indeed broken the rice bowls of numerous coolies, sedan-chair bearers, and boatmen. It was also true that the mechanized coal mine had seriously hurt many little mines and that imported goods had displaced native products."[47] From this passage it is obvious that the railroad, while cheaper and more economically efficient, destroyed many families that provided competing services. The construction of the railroad also harmed the economies of entire cities that were passed over by the new route while it benefited others along the railroad's path. Unlike the Lu River, which turned slightly south from Liling City and emptied out at Lukou, the railroad turned to the north from Liling City and entered the city of Zhuzhou. Because of this, Lukou was probably hurt by the dramatic change in the transportation route that linked it to the emerging markets including the coalfields of Pingxiang County.[48] Conversely, the railways transformed the town of Zhuzhou into one of Hunan Province's more prosperous industrial centers, since it benefited from the easy access to Pingxiang County's coal and the importance the town held as a node between the modern coalmine and the industrial center in the Wuhan Cities.

In all, the railroad between the coalmines of Anyuan and the Hanyang Ironworks greatly increased the speed and efficiency of the state-run industry and solved several old problems that hampered the initial efforts of the managers in the mining town. Even more, some small and moderate-sized cities along the route benefited by the coming of the new rail system designed to link many of China's economic centers together. The railroad system transformed the transportation economy for both the suppliers and consumers of goods, and for the transportation workers themselves. It supplanted haulers and brought Western products into the county that competed with products supplied by the local population. Therefore, not everyone greeted the railways with open arms. While some refused to purchase the items the railroad sent because they believed their usage went against their Confucian values, others feared the impact that these new consumer goods had on the local economy and their own futures.

The Managerial Center and Living Spaces: "Right royal times they had"

Just as the mining landscape was an amalgamation of Western and Chinese technology, so too the buildings that dotted the town—administrative and managerial halls, religious sites, schools, hospitals, shops, and inns—were erected by people whose ties to the town were as varied as the buildings themselves. Their architecture and function, location and usage, provide us with a tangible history of the creation of this multicultural town. The story of these buildings then is also the chronology of that community development. What is obvious from these buildings and their inhabitants is that Anyuan was a Western mining town that sat uneasily upon a Chinese landscape. The formal structures of the town of Anyuan—the managerial headquarters, hospitals, Western schools and churches—were designed to provide for the needs of the German engineers, both social and administrative, but were not entirely suitable to the culture and lifestyles of the Chinese workers.

Leinung and his assistants constructed the administrative and service buildings in the mining town of Anyuan as time and need arose. Most of the structures in the city were simple and functional. Offices staffed by German and Chinese managers were located throughout the mine site near the adits and along the foot of the mountain. While some of the buildings were Western-designed, multistoried brick buildings with verandas, most structures tended to be little more than whitewashed cottages with small porches or single-storied, nondescript structures with small windows and tiled roofs.[49] Leinung and his German associates erected offices, schools, hospitals, and other facilities necessary for the running of a modern city. At the same time, Chinese workers and commoners constructed living spaces and locations for leisure and illicit activities, while new inhabitants, both Chinese and foreign, added more cultural and religious structures to the town. Though many of the structures were simply functional, some of the buildings were significant sites that were constructed to provide a social or cultural need beyond the requirements of mining and manufacturing by one civilization or another.

The administrative center of the city was located wherever Leinung and the other Germans worked. When Leinung arrived, he initially lived in Pingxiang City but probably quickly moved into quarters closer to the mines. By 1900 more German engineers arrived with contracts that assured them adequate housing and stipulated that they serve as effective supervisors of their workers and overseers of the equipment. Over time, the engineers' expanding occupational and social needs led to the growth of a foreigner compound. Like many such compounds located throughout Asia in the late nineteenth century, the conditions were comfortable, even extravagant, and included perks that went with the job.[50] One American official named L. C. Arlington, who was stationed in the hot Hunanese provincial capital city of Changsha, traveled in 1910 to the coalmining community located in the nearby highlands, seeking cooler weather. And in his memoirs he described his visit to the German compound:

> At [Pingxiang] there was [a] colony of some forty Germans in the coal mines, and right royal times they had too, to be sure. They lived in well-furnished houses, and had a large tennis-court surrounded by electric lights, so that they could play at night. In the centre of a large garden there was a table large enough to seat some 100 people, and here tankards of German beer disappeared by the gallon, a real "*bier garten*."[51]

The buildings and houses were almost certainly part of the small neighborhood found to the south of the city along the hill near the General Horizontal Alleyway adit. From this vantage point, the German managers could look out over the entire city and keep watch over all the activities of the workers. They could also enter the mine in a moment's notice to solve whatever emergencies might arise.

While Sheng Xuanhuai assured the German engineers of adequate housing upon their arrival, petty mine bosses who hired the miners provided little more than tents or squalid housing for their employees. Since bosses hired workers through the contract labor patronage system, the labor boss was responsible for taking care of his employees. This centuries-old system generally required the boss to feed and house his workers, but these conditions were notoriously exploitative.[52] Even though the boss kept the miners on a bare subsistence lifeline, workers viewed the boss as the source of their existence and supported him accordingly. This labor devotion helped to develop relatively regimented and dedicated working gangs but also led to violence against other workers and managers and even pushed gangs into organized resistance against the foreign mangers.

After a series of labor disputes in 1904 and 1905 led by contract labor bosses seeking to undermine the authority of the foreign engineers, Leinung was forced—against his better judgment—to invest the firm's meager capital into the daily needs of the workers and build several dormitories and cafeterias to both house the workers and take direct control over labor's nonworking hours.[53] Leinung structured the buildings to house worker teams according to their status as employees. For the skilled, there were four dormitory complexes located in the east, west, north, and south of the main compound. Each dormitory building was a multistory structure that included five bedrooms with twenty-four beds. Because the workers were made of two teams—each doing twelve-hour shifts—Leinung only had to build enough dormitories to hold beds for just the half the workforce. Since one of the two teams of the workers were always in the mines, they could house twice as many workers as there were beds. In total, the structures each housed upwards of 600 workers, and therefore four dormitories provided for a total of some 2,400 men. Outside of the dormitories he erected two bathing rooms for the miners' needs. The rooms were about 30 square feet and were organized like a ship's sleeping quarters with three-tiered bunks on each side of a central aisle. They were far from comfortable, and workers complained that they were cold in the winter and were hampered by insects in the summer.[54]

In the middle of each dormitory, stood a cafeteria that the industrial investigator Gu Lang found served each miner 3 catties of rice (about 1.5 pounds) and salty vegetables per day, which he claimed were very delicious. Also, he asserted that the men took good care of the kitchen and that the workers had easy access to clean stores of foodstuffs.[55] In 1980, on the other hand, one journalist, after visiting the current government's museum in commemoration of the workers' struggles, wrote that his Chinese museum guides informed him that the dormitories housed the men like livestock and fed them gruel, only suitable for pigs and dogs.[56] Similarly, one miner remembered that all the food was purchased by the mine wholesale and was usually rotten by the time it reached the workers.[57] Since the quality of the food available for purchase from local farmers must have changed with the seasons, both statements may be true for different times of the year.

However, even though Leinung constructed the dormitories to control the miners, he still could not break the hold that the secret societies had over these men. Leinung hoped that by providing workers suitable housing and meeting their other daily needs, they would walk away from the contract labor bosses and the paternalist control of the secret societies and place their allegiance instead with him and his firm. In this endeavor, he failed: The intransigent power of the bosses continued.[58] In fact, teams of workers lived together within the dormitories, and each room where they stayed contained a sign with the name of their contract labor boss over the door. Leinung tried on at least one occasion to fire contract labor bosses and replace them with men he could trust, but the patronage power of the boss system and the influence of secret societies continued to creep into the workplace.[59]

Similarly, if Leinung hoped to control his workers in part by supplying them with food and shelter, he also eventually felt compelled to construct a medical clinic that could attend to the miners' work-related injuries. Mining is among the most dangerous forms of labor as men who entered the deep caves were far too often crushed by rock, suffocated by coal dust, or met some other untimely injury or death.[60] By one such account, thirty or forty miners died in such accidents in 1922 alone, though this official calculation is almost certainly lower than the actual number. And a more recent publication with a decidedly Communist Party bent explains that the number of deaths could be calculated as four to five miners per 10,000 tons of coal. Given that the mines produced hundreds of thousands of tons by the early twentieth century, and close to a million by the 1920s, this would suggest either that the number of deaths was in the hundreds per year or that the author was simply wildly exaggerating.[61] In either case, the mortality figures indicated the extreme danger of modern mining. And, while these accounts only refer to cave-ins and mine explosions, other problems such as diseases were equally prevalent. Even many of those fortunate enough to survive the job suffered from crippling injuries and disease due to working in the depths of the mines.[62] According to one contemporary study, over 80 percent of the surface workers and more than 90 percent of the laborers who worked beneath the surface were infected with debilitating hookworms, or "miners' anemia." Since Leinung did not see the need to construct lavatories in the mines, workers were forced to defecate anywhere they could. These unsanitary conditions fostered larvae that entered into the skin of barefoot miners. The result was often tens if not hundreds of centimeter-long worms whose teeth or jagged edges latched onto the lining of the intestine and pulled the blood out of its carriers. Infected people lost blood, protein, and iron and were too lethargic to work.[63]

In 1904, some seven years after Leinung began to expand mining in the Anyuan area, he and the other managers decided to address such problems by constructing a Western hospital at the foot of the mountain.[64] The small, white building stood alongside a cluster of office buildings and was shaded by trees where the observer Ci Fei sat to escape the heat of the day. Inside the hospital he claimed that at least one German surgeon and several tens of Chinese assistants attempted to mend the wounded miners.[65] In fact, the hospital contained most of the latest medical technology and Western-trained staff. However, even if this facility was impressive compared to the medical care available to the Chinese population as a whole, the thousands of miners who were potentially exposed to a myriad of dangers every day may have been more than the hospital could handle. Because of this the numbers of sick and injured probably overwhelmed the small clinic on a regular basis.[66]

The Western hospital also added to the cultural issues involved in the development of the mining firm. The establishment of a modern medical facility in the small isolated mining town was an improvement on the healthcare of the residents, yet it was also another facet of Western technology imposed on the lives of the Chinese inhabitants. Many Chinese were leery of Western medicine and science, fearing it involved the use of magic and other potentially harmful practices, so some workers did not trust the medicine performed in the company hospital.[67] Therefore, many miners were reluctant to take advantage of the medical facilities, turned instead to Chinese medicine for their ailments, and perhaps continued to suffer from inadequate care.

In the early 1900s while many buildings in the mining town primarily served to meet the secular needs of the foreign engineers and their Chinese workers, Chinese elites erected another building called the Zhang Zanchen Memorial Hall, which followed the Confucian notions of devotion and piety. In 1906, Zhang Zanchen, the Chinese manager of the mines and the man often said to be the founder of the modern mines, traveled to Tianjin to attend official matters. While he was there he became ill and subsequently traveled to Shanghai for medical treatment. Unfortunately, his illness was too severe, and in the early months of 1907 he passed away at the age of forty-five. Upon the announcement of his death, the Chinese managers, in a gesture of gratitude based on Chinese custom, constructed a building in his honor named the Gentleman Zhang Ancestral Hall (*Zhang gong ci*).

This memorial hall is a two-story-tall brick building that is still standing, includes verandas on the façades of both wings of the building, and contains several classrooms on both floors.[68] When it was first completed it was used in part to educate local elites who sent their sons to schools to prepare for the civil service exams or to achieve status as merchants or officials in the Confucian order. Not long after its completion, however, the school fell into financial difficulties and was subsequently taken over by British missionaries who acted as the deans if not the teachers. Under the control of the foreign clergy and Leinung, the school focused on Christian teachings and Western mining skills rather than Confucian ethics or Chinese government. The Department of Economics courses, in particular, emphasized mining and management. One Western observer wrote of the professional training,

> An instance of the broad-minded and commendable attitude of the engineers in charge, their devotion to the cause, and desire to help the Chinese in every particular, is in in the recent establishment of a school of mines for the purpose of training Chinese in the profession of mining engineering. The school

> is already equipped with the necessary apparatus and has all the requirement found in the ordinary college course with the exception that the theoretical subjects are not pursued so far. A three-years preparatory course is required before students are allowed to enter the mining course which continues from 4 to 5 years, the first being devoted to actual work, and study in practical mining and allied work. This is the policy now generally pursued in the technical colleges of Germany. The students are to be given every facility for supplementing the class-room work with the study of the practical methods as carried out daily in the mine, and vacation months must be spent in actual work about the mine and the company will supply and allot the positions. There are now over 60 students in the preparatory course.[69]

This school, which was referred to as a "public school" in the gazetteer, was organized into three levels based on ability. It reportedly educated about 270 to 280 boys and some girls. And unlike schools of the past that served gentry families almost exclusively, the mine firm provided miners' children with special compensation for attending the schools. These families, who in the past had no access to quality education, made up as much as 30 percent of the student body. Ci Fei, who was a visiting Liling County teacher, observed a class taught by a teacher from Jiangsu Province. He determined that the class was poorly taught, though it was not entirely without merit.[70] The history of this building thus presents in tangible form the notion that foreign dominion literally replaced Chinese rule.

Along with the school's Western classes, the Episcopal church delivered Western culture and ideology into the mining town. The church was started by a Chinese Episcopal minister determined to spread the word of the Christian faith. In 1900 he arrived in the county seat and began proselytizing. There he found little interest in his message among the Chinese of the gentry-dominated city and so he moved on to the mining town, where he thought the foreign engineers and Western-educated Chinese elites might be more hospitable and attentive. By 1915, the minister secured funding for a major land purchase in the mining town, allowing him to situate his church near the General Horizontal Alleyway among the lower hills surrounding Tianzi looking down on the entire mining town. Once he established the church, he ministered to the European engineers and to some skilled Chinese engineers who were from the Western provinces of China and had converted to Christianity.[71]

The history of this church is important for at least two reasons. First, it indicates in yet another fashion that the implementation of the mechanized mines changed the architecture of the town even as it altered the mining process. Foreigners and Western-influenced Chinese who lived in the town required services including religious teachings that the local population could not provide. These services were provided in a tangible manner in the form of a Christian church nestled in the hills of a Confucian and Buddhist community. Second, the church added yet another source of contention between the local population and the outsiders. As early as summer of 1896, when students heard that a foreign engineer was to arrive in Pingxiang County and investigate the local coal deposits for possible modernization, they complained that the foreigner's arrival would almost certainly spark Christian missionary work in the county, something they certainly abhorred. To calm the fears of these men, Gu Jiaxiang, the county magistrate, argued that no such interference with Western religion would happen. He simply suggested that the foreigner was coming to work toward increasing coal output and nothing more.[72] As will be discussed in more detail in chapter 6, Gu proved to be seriously mistaken about the coming of Christianity as well as many other social consequences of modern mining coming in the county.

If the establishment of a Western school and Episcopal church indicated the cultural invasion of Western thinking, then participation in both of those institutions by Chinese gentry and higher-class Chinese workers suggested their real impact. Elite Chinese families and newly emerging professionals in the mining community sent their children to a new Western school rather than the old Confucian academies and were attending the new Christian church rather than, or at least along with, the religious shrines of their ancestors. These edifices of Western culture and intellect sought to replace the old ways of thinking and even the basis of China's social order in this community.

And if the new Christian church sat atop a hill overlooking the city as a symbolic expression of Western power and influence, similarly the new managerial hall was physically even more imposing and its symbolism could not be more obvious. Laborers completed the building in 1916, the year Sheng Xuanhuai died, and so they called it the Sheng Xuanhuai Memorial Hall. It was located on a hill just above the church and signified in dramatic form the power of the managers, foreign and Chinese alike.[73] Unlike most Chinese government offices, the stone structure was multilevel. Also, very different from yamen in county seats that were located in the central or northern section of the city and faced southward by mandate, Leinung located the Sheng Xuanhuai Memorial Hall near the General Horizontal Alleyway looking over the town south of the city and facing north.[74] Like many bungalows built by Western imperialists of the time, the memorial hall was in a neoclassical design with plenty of open patio space on the lower floors but with a decorated Chinese-type roof adorning the top.[75] While the structure signified the strength of the foreign managers who oversaw the mines, the tiled and slightly concave-shaped roof also indicated an acceptance of its Chinese surroundings. In this way, the architecture of the city was both Western and Chinese and followed the local culture and customs even while imposing a foreign order.

In my own trip to the mines, I entered into the Sheng Xuanhuai Memorial Hall to glimpse the inner layout of the structure. When I entered the front door of the headquarters, I was immediately standing on a wooden floor inside a small visiting area. After further passing through the initial greeting space, I was surrounded by offices, internal stairways, and attached balconies that connect to several rooms on each level. On the outside, the building followed the classic bungalow design with verandas that surrounded the building's intricate brick and plaster-cast façades that allowed the inhabitants to spend their days in the shade even in the heat of the south China summers.[76] Attached to the back of the main building stood a large two-story attached shed with no interior walls. The façade of the shed followed the designs of the main building and was large enough for storage.

The foreign and Chinese managers clearly designed and located this administrative building to portray power and authority over the workers and Chinese inhabitants of the town. However, this power came from several locations. While Leinung and the other German engineers likely constructed it for their own housing, in fact, by the last half of the 1910s Chinese engineers and managers filled the positions of some of the Westerners. Moreover, since the hall's architecture was comprised of Western functionalism and a type of pseudo-Chinese aesthetic, it would seem that the designers were themselves willing to integrate multiple voices into the building's statement of power. Finally, by naming the building after Sheng Xuanhuai, the German managers continued to accept ultimate Chinese authority at least in word if not in deed.

As the foreigners accepted final authority of the Westernized mines to the Chinese officials, so too the Chinese infused their own culture on the modern town with local markets, inns, and tea shops as well as such "illegal spaces" as gambling halls and brothels.[77] For example, some local merchants and families infused Chinese culture when they moved into the town looking for work and a place to live. An even more important source of architectural change in Anyuan was the secret society called the Gelaohui. This antigovernment secret society oversaw the labor guild and ran, or had influence in, many of the other business establishments. In his book on the early life of Mao Zedong, Li Jui writes, "The foremen, overseers, and the bosses of the local secret societies also jointly set up gambling halls, opium dens, houses of prostitution etc., in the neighborhood of the mines."[78] This quote suggests that the Chinese created these shops under the noses of the German engineers and within the confines of the mining town. These spaces were very popular, as sources describe extensive usage of opium and reliance on prostitution among the workers.[79] More importantly, the Gelaohui's membership gathered regularly for both meetings and illicit behavior and may have included commoner and low-level gentry figures alike.[80] As will be discussed in the last chapter, by the first years of the twentieth century, the Gelaohui organized thousands of members in the Anyuan area and many thousands more throughout Pingxiang County.

The miners' social needs were not always met in this enclosed valley, however. For example, even as many Chinese turned to other public spaces for recreation, travelers mentioned only uninviting small parks, poorly suited for workers who hoped to escape the loud and hot mining town.[81] Also, like most Chinese cities, Anyuan did not have a cemetery for its inhabitants. People who lived in Anyuan as coalminers or factory workers were legally inhabitants of some other location and therefore not buried in the vicinity. The dead had to be taken to their home villages for burial. One worker remembered seeing bodies lying in a building awaiting movement for burial. The building was not a funeral hall or place of mourning. Rather, it was a warehouse containing machinery and other equipment owned by the Mine Bureau.[82] No doubt as mothers and wives sent their men off to the mine, they did so hoping that their labor would provide supplemental income to the family. For so many, however, what they received instead were the mangled bodies of their loved ones.

Overall, the Germans designed the formal structures of the town of Anyuan to provide for the needs of the engineers—both social and administrative—but those structures were not entirely suitable to the culture and lifestyles of the Chinese workers. Buildings included ancestral halls and Chinese secret society cites within view of a Christian church and a Western-designed managerial headquarters. The foreign compound held a beer garden; the Chinese streets provided workers with tea shops, opium dens, and halls of prostitution. While basic medical treatment was provided, care was at times inadequate and could not prevent men from dying in the mines. Moreover, when death came—as it all too often did—the town's facilities shipped the bodies out while the foreigners and Chinese bosses recruited new workers to take their place.

Labor Bosses and Labor Life

The establishment of the mechanized mines affected the landscape and it also brought the labor force into new relationships with their superiors. Rather than building patron-client relations between peasants and landlords, coolies and gentry, the new system joined together more than ten thousand laborers from several counties and provinces with a new elite of patriarchal contract labor bosses who in turn established contractual relations with official-merchants and foreigners. Controlling mine labor had two especially significant effects. First, Sheng Xuanhuai's appointees hired thousands of laborers to work in the modernized mines under their management. The firm subjected the men and young boys to labor conditions and wages instituted by the Western managers to make the process more efficient and technologically advanced. Second, contract labor bosses employed patronage relations with their hires that formed a buffer between labor and management and stifled much of the modernization and efficiency that Western technology was designed to instill.

Management control over labor was a gradual process. From the first years of the modernization process in Pingxiang County's mines, Chinese and foreigners negotiated over conditions and pay. In 1896, the Wen leaders of the Guangtaifu Lineage Trust admitted that many of their laborers and boat haulers took advantage of their employers by diluting their coal output with shale. They argued that mine workers required direct oversight and promised to hire men to assure the quality of the output. However, oversight of an operation, even one as small as the initial workings of the Guangtaifu Lineage Trust, required more managers than the firm could afford. To offset this problem, they enacted a set of rules that forced all the employees to evaluate each other and posted these requirements on a stele in the early summer of 1897 for all the workers to see. In this lengthy passage, the managers laid out each of the tasks the workers performed and the means they used to oversee their fellow employees:

> We must charge the picker in the mine first to take the pit from the top wall layer and pick holes and remove and clean it way completely. Then dig out the coal. Before extracting from the

> pit, cause the workers who pick up the coal to subsequently sieve it through and meticulously examine [to make sure there is] no rock, then send it out of the pit to be sifted. Again cause the sifter to examine for impurities utilizing a fine screen to sift it clean. Then again cause the coke oven operator to pick up the coal to examine it and clean it for impurities; only then can it be put into coking ovens. In this way we can rectify [the process] from the beginning to get a pure source then we will be able to process fine coke increasing the circulation for marketing. If diggers scheme to mix in the imperfections into the coal, as soon as the hauler examines it, resolve to punish the digger today's wages and reward the hauler. If the hauler is not careful and attentive as soon as it is screened by the sifter and has imperfections, resolve to punish the hauler today's wages and reward the sifter. [If] the sifter sifts improperly, as soon as the smelter examines it and it has imperfections, punish the sifter one day's wages and reward the smelter. After the coke comes out of the oven [if] there are imperfections then punish the smelter three days wages and reward the other workers to show your intentions.[83]

This pronouncement describes the division of labor and exactions inflicted upon the workers. The mining process was increasingly complex, yet the lineages in Pingxiang County continued to utilize simple labor oversight and management systems that dated back for centuries. The lineage leaders chose not to expend greater investments hiring skilled managers and workers but instead simply brought in local workers and gave them simple jobs with simple technology. For those workers who performed their duties poorly or engaged in various forms of sabotage or corruption, the Wen lineage used financial penalties to force them into compliance.

In addition to fines and other similar penalties for substandard performances, lineage leaders turned to the same violence and beatings that gentry used in the villages for centuries to maintain discipline. Moreover, once the mines came under the authority of Zhang Zanchen and Gustav Leinung, violence was commonly incorporated as a means of controlling labor. For instance, the local gentry wrote a petition accusing Lu Hongchang of various misdeeds, his horrible temper, and his tendency to beat local inhabitants for any reason he saw fit.[84] Even after Sheng Xuanhuai relieved Lu of command over the mine enterprise, he remained for several years as an assistant to Zhang Zanchen, and no doubt his treatment of the workers continued to be brutal.[85] Similarly, Chinese workers complained vociferously that Leinung and his German associates used various forms of violence to control their workers. Based on the findings of Communist Party leaders, historian Lynda Shaffer explains that

> The staff, both Chinese and foreign, and the contractors were allowed to beat the workers or otherwise punish them, and the mining authorities did not interfere, since this "contributed to their authority in dealing with the workers." Among the variety of punishments available to the staff and contractors were making the worker kneel on a hot stove, or making him wear a cangue around his neck. Other possibilities included forcing the worker to carry a steel ball on his back, horsewhipping or turning the worker over to the police to be interrogated by torture.[86]

The treatment of workers in the mines remained brutal and oppressive. For instance, one young boy named Yuan Fawu remembered being physically assaulted on numerous occasions by German and Chinese managers as well as seeing others beaten while he was working inside the mines. Police and soldiers routinely put down labor unrest. Oftentimes, these men used violence and even deadly force and mass executions to quell the workers.[87] More commonly, the continued harassment of miners by Chinese and foreigners alike fueled greater animosity between managers and workers and crippled the labor discipline and efficiency that the beatings were supposed to ensure.

Beyond sheer brutality, to improve output while increasing the size and scope of the mechanized mines, Gustav Leinung had to recruit more and better Chinese laborers than had ever worked in the county's mines. He began by simply employing local workers from the mines to continue their efforts as they had for generations. However, he was not satisfied with the miners' performance, and he complained that many of them were addicted to opium and refused to "work regular hours."[88] His complaints about the miners' working habits may not have been unfounded. Addiction to opium was likely widespread throughout the highlands where opium dens were prevalent. It may also be true that the Pingxiang County miners simply did not work hard enough to satisfy his performance demands. The German managers implied that they were used to a higher level of skill among German miners and that comparatively the Chinese miners employed poor work habits.[89]

In order to solve these problems, Leinung decided he would have to hire workers from outside Pingxiang County, further undercutting the hold that the local population had on the mines. This decision was fraught with political consequences since, as has been clearly shown earlier, even Sheng Xuanhuai's decision to bring in German engineers led to anger and conflict.[90] To this end, Leinung hired miners from other counties and mines in the region. Unlike the men who had worked the mines before, these new recruits had no connections to the Wen lineage or any other powerful family in the county. Not only had the firm taken the land from the powerful families but control of labor power had as well.

Because recruitment in China could not be accomplished through open labor markets, the mining firm utilized the centuries-old contract labor system that incorporated Confucian-styled paternalism with the ability to bring together large numbers of skilled and unskilled workers. The majority of the new workers came from Leiyang County in southern Hunan Province and Daye County, Hubei Province.[91] These men probably arrived at the mines as skilled, semi-skilled, and unskilled mine workers, with various backgrounds in premodern mining methods, factory labor, or transportation work. While in 1896 the vast majority of the miners in the area were Pingxiang County residents, by 1899 half of all the miners were from Hunan Province.[92] There are several reasons why Leinung sought Hunanese miners. First, Hunan was nearby and so the workers could be recruited and brought to the Pingxiang Coalmines easily. Second, the Leiyang County highlands held rich deposits of anthracite coal mined for generations and thus the region contained a large number of experienced coalminers.[93] The workers from Hubei Province, on the other hand, primarily arrived from the coal and iron mines at Daye, where Leinung had previously worked as one of Sheng Xuanhuai's engineers. He hired people there whom he felt he could trust and whose work he had already observed.[94] Finally, Leinung also recruited some Chinese miners from the east coast who were skilled in Western technology. These men were both better trained and more fully integrated into Western science and culture. Unfortunately for Leinung, however, these recruits were too few in number to overcome the overall poor working performance of the local laborers.[95]

In fact, Leinung's hope of an effective labor force never materialized. Even as he brought in people from outside of Pingxiang County to work for him, these laborers simultaneously joined secret society–based contract labor gangs that remained insulated from the direct oversight of the German managers. Bosses tended to recruit from within their own natal community and therefore the workers often viewed themselves as *tongxiang*, that is, members of the same township. Moreover, once they arrived at the mining town, their gang bosses recruited them into the local secret society, the Gelaohui, which was itself a patronage organization that gained its strength from political and illicit economic ties throughout the countryside.[96] Leinung explained that when he first recruited workers they were all members of the society. To undermine Gelaohui authority he fired these bosses and hired new ones from among the best workers. Unfortunately for him, the Gelaohui quickly recruited these new bosses, and they in turn brought their workers into the secret society organization as well.[97] These associations often competed with each other in the mines as members of various gangs fought for the best jobs or the most lucrative contracts. Contract labor bosses and workers engaged in fistfights to determine who received the best working conditions. Bosses that were most successful in these skirmishes in turn received better labor recruits.[98] In this manner, workers were organized along premodern systems that kept them apart from management but also undermined common class unity as well.

Descriptions of individuals recruited into the mines are sketchy but tend to support the claim that connections played important roles in securing employment. The miner Cai Shufan explained that his family originally came from the Daye Iron Ore Mines. Though his grandfather was a peasant, his father and uncle worked as miners and factory workers in the Daye Iron Ore Mines and the Hanyang Ironworks before the firm recruited them to work in Pingxiang County. Cai explained that three days after he was born in 1908, his father, mother, uncle, and possibly other family members migrated to Anyuan. When Cai was growing up, his father could not provide for the whole family on his miner's wages. Therefore, at the age of fifteen, he went to work as a laborer in the machine buildings.[99]

The recruiting process is not fully described in Cai's biography. The stories he provided alluded to at least two forms of labor recruitment that are both part of the premodern Chinese patronage system. First, contract labor bosses recruited Cai Shufan's family from the Daye Iron Ore Mines to work in the Pingxiang County Coalmines. They had worked in the factories and mines in Hubei Province for at least several years when the recruiters hired them to work in Pingxiang County. Second, after Cai Shufan's family worked in Pingxiang County for about fifteen years, his father used his influence to get him hired on by a contract labor boss. Friends and relatives who agreed to vouch for a worker facilitated recruitment as well. In this way, groups of people tied by any number of relationships formed labor teams.[100]

In some cases, impoverished people migrated to Anyuan or its supporting facilities hoping to find full-time employment, but even then employment was accomplished through contacts with labor gangs. In his memoir Yuan Fawu tells of being from a peasant family in a small Hunanese village. His father died from the hardships of peasant life when Yuan was a little boy. Without an adult male to help provide for the family, his mother decided to leave the fields. Fellow villagers told them that there were jobs available in the coalmines in Pingxiang County and so he, his mother, and his two younger sisters took what they could carry on their backs and walked up the mountains to the town of Anyuan. After working for a short period collecting coal on the roads and selling it to the mine, a friendly miner introduced Yuan to a contract labor boss. The boss hired him to work in the mines where he did several jobs and worked in more than one location with the boss's team of laborers.[101]

It is also true that itinerant laborers gained employment for brief periods of time. Before becoming a member of the Communist Party and eventually the army chief of staff, Yang Dezhi left his home in Liling County to work in the Pingxiang County Coalmines. He was born in 1910 and worked as a cowherd for a local landlord from 1921 to 1924. Then in 1924, Yang worked for a half year in Anyuan before returning to his home county where he again worked for the same landlord. While it is unclear how he gained employment in the mines, since he was living in neighboring Liling County prior to his brief tenure at Anyuan, it is likely that he was able to make contacts that secured him a job. In any case, his term was brief, no doubt like many workers who turned to the mines in times of desperation.[102]

These stories clearly point out that mining was primarily done by young men who worked long hours inside the mine shafts, doubly exploited by contract labor bosses as well as the Western engineers. When the work shift began with the sound of a whistle at 6:00 a.m., half of the estimated thirteen thousand workers began their twelve-hour shift, and at 6:00 p.m. the other half began theirs.[103] Most of the miners were slightly over twenty years old, and all were males. About 80 percent of the workers labored inside the mines. This included virtually all the skilled miners and many of the unskilled laborers as well. Those who worked outside the mines included about two thousand workers who coked the coal and some of the nearly one thousand machinists as well as several hundred "coolie" laborers who moved the coal into and out of the factories and trains. These last men reportedly transported as much as 2,000 tons of coal per day at such cheap rates that much of the machinery found in contemporary Western mines was unnecessary.[104]

Because these workers were vulnerable and politically powerless, low wages provided by the mining firm and the contract labor bosses represented another form of exploitation. While on average Chinese miners made no more than thirty-five cents Mexican silver per day, the actual pay from worker to worker was often less than that because most workers were taken advantage of by the contract labor bosses. Under the contractual relations with the mine manager, the boss received all the wages of the work team and then divided up the pay however he saw fit. Upon receipt of the money, the boss usually took at least 10 percent for his efforts. In some cases, he took more, often reaching nearly 50 percent of all the income of the entire group in payment for board, clothing, and other incidentals.[105] Moreover, the pay scale often varied based on several different scales and measures including the age and skill levels of the workers.[106] While it is likely true that the miners' wages were slightly higher than peasant incomes, the hazards of mining often shortened the life of the worker, and working in mines required the miner to separate from the kinship ties and family property that traditionally provided for survival during sickness and old age. In this way, mining was taken up by desperate peasants rather than those who sought higher pay or avenues to success.

Experiencing so many problems with Chinese workers, even poorly paid ones, Leinung argued that the mines would be more productive of the firm hired more Western engineers and miners to work the mines and to train the Chinese to work them as well. Sheng Xuanhuai, on the other hand, continued to oppose the hiring of more Westerners, though he begrudgingly accepted Leinung's request for more Western engineers. On March 23, 1899, the firm drew up a standard contract for all foreign laborers at the Pingxiang County Coalmines delineating the requirements of service and the agreed-upon accommodations by the company. Under the three-year contract, the firm provided the men with a ticket on a steamer to China and their pay of £40 per month began upon getting on the ship. When they arrived, they were given a place to live and a bank account and were assured of medical treatment and other benefits while they were responsible for the maintenance of their equipment, the training of their Chinese workers, and were to perform as commanded by Leinung, the chief manager of the mines.[107] By the mid-1910s as many as forty German engineers and associate staff were on the payroll in the Pingxiang County coalmines under Leinung's authority.[108]

The increased numbers of foreign experts did not, however, lead to a more efficient labor force or more complete transformation of the production process. Training and development of skilled labor was spotty at best. For example, when one observer asked a group of miners how their tools worked, they explained that they were not clear regarding some of the specifics. Indeed, Westerners often complained that the Chinese failed to understand their commands or follow their orders. While it may be true that in some cases workers intentionally resisted learning from the Westerners or adapting Western techniques, it would seem that even years after the implementation of the machines, foreign engineers failed to train their men adequately.[109]

Furthermore, no matter how hard the Chinese men worked, they were not able to secure lucrative salaries enjoyed by their German counterparts. While laborers such as the railroad worker Wang Zheng, who was discussed earlier, indicated that greater expertise led to higher pay grades, Sheng Xuanhuai's industrial enterprise paid miners and railroad workers far less than the Chinese managers and European engineers and service people. In his contract dated June 5, 1905, Lu Hongchang was given a base pay of 200 taels per month and was promised as much as an additional 200 taels per month if he could keep his operating expenses down.[110] While this salary is much higher than most Chinese workers received, it was still far less than the £40 monthly salary paid to the German workers. Given their pay and the relative economic conditions in Anyuan, the German engineers lived quite comfortably. But, most Chinese miners' wages remained ruinously low, and these men saw no hope of advancement in the mines.

At the same time, because mining wages were slightly better than the incomes peasants could hope to make selling their crops, some families sent one or more of their members to work in the mines and factories to supplement their peasant incomes. The railroad worker Wang Zheng explained that his father left the family for employment in Changsha while he took another job in Liuyang. Initially, they both planned to send money home to help pay for the family's needs. However, with the poor pay scales awarded to laborers, it soon became obvious that neither could afford to provide extra income and the women and children were forced to fend for themselves. Wang explains that his mother "had to sell my young brother to a (landlord) family as a slave in the fields, receiving three hundred dollars for him."[111] This further pushes the point that the mines and factories were paying their workers barely more than the minimum required for their own subsistence. For the rest of the family to survive, women and children worked the fields or themselves turned to the towns and factories for supplemental incomes.

In fact, to supplement the daily needs of the mine workers, local women and children took various jobs in Anyuan performing tasks usually held by women family members in the villages. The production scheme in the town of Anyuan required thousands of men to work twelve-hour days for the entire calendar year while living far from their lineages and homes. This meant that daily needs such as food preparation and clothing procurement and even sexual gratification were often done by someone else for pay rather than as part of the portfolio of tasks and jobs completed by one's family. People who in the past would have provided these services for their family members in their homes and villages instead gained employment as cooks, maids, food sellers, or other similar jobs. In essence, the coming of the mechanized mine and the expansion of the money economy broke up the tasks of the family into a series of jobs that people did for pay. This change, therefore, had not only economic but social ramifications as well.[112]

Many men and women worked in Anyuan as store owners and clerks. These shops and stores sold goods and services to workers to supplement their daily needs. Along with the brothels and gambling houses described earlier, one observer wrote that there was a large vegetable market on the Dama road in the town of Anyuan. This market had among other things an abundance of pigs fattened on grains. Also at least one street contained over one hundred shop fronts. Both of these markets were located near the train station where goods were brought from Wuhan, Shanghai, and foreign distributors into the town of Anyuan and coal transported to the Hanyang Ironworks.[113]

Similarly, the higher-paid foreign and Chinese engineers hired many personal assistants from among the migrant or peasant population in the community. For example, Leinung hired assistants to do a number of different activities. In an account from 1916, a "servant" informed Leinung that someone posted a threatening sign on his front door.[114] Also, the description of a German beer hall in the foreign compound certainly brings to mind the need for wait staff and bartenders.[115] Though these domestic laborers were probably commoners hired specifically for the purposes of supplying the German managers with the comforts of home, the fact that at least one of them could read indicates that they may have been more than just peasants-turned-laborers. As is seen in so many histories of plantations or urban life, the servants and indoor workers gained greater access to elites and their families and often enjoyed their work—poorly paid though they were—more than the jobs of the factory workers and miners outside. These jobs gave the employees different social status from their laboring counterparts and further complicated the social structure of the community.

An even more complete description of non-miner labor strategies comes from the recollections of Yuan Fawu. He recalled that he and his family engaged in several jobs including collecting and selling pieces of coal. After Yuan was hired into the mines, his mother and sisters continued to collect coal. Yuan's mother and sisters also picked up work doing laundry for miners. Income from these jobs allowed the family to survive in their home after Yuan became temporarily disabled in the mine.[116] Similarly, Nishizawa Kimio, a Japanese engineer who toured all the mines and factories attached to the Hanyang Ironworks in the early 1910s, observed that family members often engaged in various forms of employment to support the miners' incomes. According to him:

> The day laborers, or those having their own homes, mostly live with their families, which, on the average, consist each of three members besides the wage earner. They each support themselves and their families at about twenty *sen* a day, which they are paid as their wage. They generally have extra incomes in the form of earnings by their women folks and children which are more than sufficient to meet disbursements on clothing and other occasional expenses.[117]

Nishizawa suggests here that rather than leaving their families behind, some of the better-paid miners maintained their families in Anyuan as long as the other family members took supplemental jobs in the town. These families left the villages behind, turned their backs on subsistence farming, and became urbanites. And even though their status changed from peasant to worker, the fact that they made this leap tells us that they were still able to live at a financial level equivalent or better than that back home. This decision also allowed families to stay together rather than be forced apart under the stress of the modern industrial scheme.

For those who did migrate to the Pingxiang County Coalmines without their spouses, one added expense may well have been for sexual contact. The ratio of men to women in Anyuan was no doubt extremely high given the town's rapid growth and focus on a gender-segregated occupation.[118] Miners were commonly described as a raucous lot who looked for many forms of entertainment and diversion from the horrors and tediousness of their jobs. Some who could not turn to their family's warmth after a day's work sought instead the comfort of sexual companionship and gratification. In one document by a Communist Party cadre dated 1924 the author explained, "[Workers'] lives are extremely agonizing, those who can marry are very few, whoring and gambling are these peoples' ordinary habit."[119] The statement makes clear that higher-paid workers gained the added benefit of a wife and family in the town. However, the vast majority of the low-paid workers in the mines and factories found entertainment and the companionship of women more frequently in gambling houses and bordellos.[120]

This discussion about prostitution is perhaps more speculative than real, and the depiction of all poor boys turning to prostitutes may be too simplistic. Mine worker Yuan Fawu explained that even though he was an unskilled worker, he was still able to marry and have a family. His mother decided that the money she and her daughters collected as laundry workers, combined with the increased wages Yuan received after a successful strike in 1924, was sufficient for him to start a family. His mother arranged for him to marry a young neighbor girl who came from an equally poor household facing many of the same hardships that Yuan and his family experienced.[121] Based on this description, it appears that marriage was not out of the question for at least some of the poorer workers. Rather, it is likely that the presence of Yuan Fawu's family to provide supplemental income in the mine town greatly improved the chances of marriage for the mine worker. Not only was the family in this way able to survive without experiencing many of the financial burdens of peasant life, but also Yuan Fawu's mother no doubt watched over him and kept him from some of the harmful practices common among miners.

As miners were organized by contract labor bosses, frequented secret society–operated gambling halls and brothels, and even at times maintained loose connections with their lineages and families, they were not strictly under the control of the mining firm. The expansion of industrial labor in Pingxiang County did not lead to the creation of a fully regimented labor-management system, rather premodern and local social and political institutions bifurcated these relationships. In the years prior to mechanization, workers maintained strong ties to their lineages and community as families and villages recruited workers from within their natal areas. In order to break this bond, Leinung turned to outsiders to work the mines. However, this in and of itself did not undermine the hold of the local communities. Instead, the lack of a modern labor market forced the mines to turn to patron-client-styled contract labor systems. Even as

Leinung recruited miners from outside of Pingxiang County, these gangs of workers were often made up of relatives and neighbors and had lineage or community ties to the contract boss who hired and trained them. Moreover, the Anyuan secret societies quickly recruited or coerced the new contract labor bosses into their association. The integration of the Gelaohui with local opium dens, prostitution, and gambling houses kept the workers connected to the social organizations more closely linked to the Chinese world than the emerging modern one overseen by foreign leaders.

The Lows and Very Lows of Modernization: The Fall of the Hanyeping Coal and Iron Company, Incorporated

While modernization of coalmining in Pingxiang County was progressing apace, the larger industrial scheme as established by Zhang Zhidong and Sheng Xuanhuai continued to experience ups and downs that ultimately led to its ruin even as the empire itself was in the final stage of collapse. Because the firm's links to the court was a patronage web of political satrapies and financial safety nets, the fall of the state left the firm with no legitimacy or means of support. The effects of the firm's economic failure bankrupted the factories in Hanyang and forced the managers to send Daye County's iron ore to Japan as payment for their debts rather that to the factories for further Chinese manufacturing. In Pingxiang County, corporate failure meant that the firm could no longer provide technological upgrades or employ trained engineers, while mine officials and county officials lost their authority to maintain order within the mine and in the county as a whole. Under this stress, the process of technological transfer for the mines did not continue to its completion and this led directly to their demise.

The modernization of Pingxiang County's coalmines began because the Huguang governor-general envisioned an integrated industrial firm with the central ironworks needed to construct a railroad linking Beijing to Guangzhou. To support this factory, Zhang purchased an iron ore mine and began the process of securing a stable source of coal and coke. The three individual mechanized firms acted as a single industrial project to make iron and steel that, in turn, supplied metals for railroad rails, as well as military and civilian manufacturing needs. And by all accounts, through mechanization the firm was able to increase output in mining and manufacturing significantly. The Pingxiang County Coalmines alone raised coal output a hundredfold between 1898 when raw mineral deliveries to the ironworks were calculated at 10,000 tons and well over a million by the eve of the 1911 Revolution.[122]

However, by the early 1900s, the costs of production began dramatically straining the imperial coffers. After an initial $3 million loan from Germany by 1898, the Japanese government provided a slightly smaller one in 1903 in exchange for iron ore as well as cash, a deal that quickly benefited the Japanese as they almost immediately used the funds and resources to attack Russia at Port Arthur. More loans from Japanese banks soon followed. By 1906, Lu Hongchang calculated that the Pingxiang County Coalmines alone owed in excess of 2 million taels to all its creditors.[123]

With the debts piling up for the imperial government, Sheng Xuanhuai used his political and economic leverage to take control of the entire industrial project on his own as a single company collectively named the Hanyeping Coal and Iron Company, Incorporated. The contract for sale, drawn up on November 4, 1907, and signed by Sheng, agreed to take control of all assets of the company and to pay off all the stockholders. A subsequent petition to the Bureau of Industry and Agriculture, submitted in March 1908, officially declared the new corporation of the Hanyeping Coal and Iron Company, Incorporated. Under the new organization the firm became a private corporation run by the merchant Sheng Xuanhuai rather than a *guandu shangban* firm as it had been since the now-deceased Zhang Zhidong established it in the 1890s.[124] Sheng thus instituted merchant management, but even the Chinese merchants and reform-minded managers failed to make a difference. He eventually took out further loans to pay for the first ones. In all, he took out no fewer than eleven loans from Japanese banks after 1907 for a total of around ¥30 million at relatively high interest rates of 7 to 8.5 percent.[125] In fact on the eve of the emperor's final abdication, Sheng took out loans from the Yokohama Specie bank in Japan, while he also set up business agreements with American and Belgian engineering firms to purchase all the machinery and pay for the labor and other costs of this large undertaking.[126] In all, the history of the Hanyeping Coal and Iron Company, Incorporated, during the remainder of the Qing dynastic period is a story of Sheng's attempt to stay one step ahead of his creditors.

But it was not only economic problems that undermined the corporation's hopes of success but political ones as well. In the last months of 1911 the Qing dynasty fell, and with it the legitimacy of Hanyeping's most important leader, Sheng Xuanhuai. Revolutionaries denounced Sheng as a traitor and he was briefly forced to flee for his life. And the unrest of the revolution, which began in the Wuhan Cities, forced the industry leaders to shut down the blast furnaces and required more capital to repair and refire them again.[127] However, even as Sheng and his associates attempted to get back on their feet, the Japanese continued to press the Chinese leaders to hand over their minerals or even to relinquish control of the Hanyeping Company in payment of the loans Sheng had incurred. The industrialist and official F. R. Tegengren writes that the Japanese took "the fullest possible advantage of the embarrassment of the company, by securing from it at the lowest price the raw materials—iron ore and pig iron—that their own industry was in urgent need of, leaving the company little or no profit or even causing it a direct loss."[128] Upon the abdication of the Chinese emperor in early 1912, the Japanese hoped to gain partial control of Hanyeping through a joint-ownership arrangement. Unfortunately for Japan, when the military leader Yuan Shikai assumed the presidency, he used his authority to attempt to nationalize all the empire's iron ore and the Hanyeping Company as well.[129] Yet even as he was trying to take control of the firm, it lost money every year, with annual losses totaling anywhere from $100,000 to almost $2.9 million.[130] However, Yuan was not deterred because it was obvious that the value of the firm to the nascent republic was greater than the individual parts, and so the Chinese sought to hold on to it as a component of its national sovereignty.

At the same time, the Japanese industrial and political leaders were aware of the value of the ironworks to both China and their own future, and they sought to secure them from Yuan's administration. Even as Japan was experiencing relatively strong economic growth, particularly from such modern industrial schemes as its own Mitusi-based Yawata Ironworks, it lacked the natural resources that were found in Daye and Pingxiang counties. To address this problem, Japanese loans to Sheng Xuanhuai were in fact to be paid with the iron ore Japan desperately lacked. Japanese investments had forced the Chinese to agree to deliver 15 million tons worth of iron ore and 8 million tons of pig iron over the next forty years, deliveries almost certainly beyond the resources of the county's mines.[131] Sheng Xuanhuai apparently agreed to these loans in part because he had been told the resources available to him were sufficient. Leinung's estimates regarding Daye County's iron ore deposits were almost certainly overly optimistic given the current level of technology the Chinese and German engineers employed at the time. Two Western engineers and one Chinese manager of the Daye County mines reduced his estimates by a third or more, suggesting that Leinung's figures were dramatically optimistic.[132]

More importantly, Leinung argued that Pingxiang County's coal deposits were sufficient as the main source of bituminous coal for smelting the iron ore being mined in Daye County. In words similar to those used to describe Daye County's iron ore deposits, Leinung argued repeatedly that the county's high quality deposits would last "several hundred years with an output of 1,000,000 tons annually."[133] Once again, Leinung's estimates appear to be far too optimistic since the coalfields were high in moisture as well as ash, which therefore raised the cost and lowered the quality of the metals smelted in the Hanyang Ironworks.[134] Furthermore, even after engineers completed the construction of the railroad line all the way to the Wuhan Cities, the costs of transportation to Daye County remained excessive and harmed the production scheme.[135] Perhaps the most important problem with Leinung's assumptions is that the coalfields he claimed would produce minerals for hundreds of years are essentially gone just one hundred or so years after his tests were completed.[136]

Because the entire process failed to produce the quantity and quality of purified mineral that the Chinese had hoped for or that the German engineers had promised, Sheng was faced with endless debts he could never pay back. And based on these debts, the Mitsui Corporation as well as the Japanese government hoped to use these loans as leverage that would allow them to take direct control of the mines.[137] It was this set of circumstances that led the Japanese to list joint control of the Hanyeping Company as one of the terms of the Twenty-One Demands in 1915. While the demands called for China to accept Japanese imperialism and occupation of Chinese territories, they also stipulated that China accept closer cooperation with Japan regarding the Hanyeping Company. Group 3 of the demands reads in part:

> Article #1—The two contracting parties mutually agree that when the opportune moment arrives the Hanyeping Company shall be made a joint concern of the two nations, and that they further agree that without the previous consent of Japan, China shall not by her own act dispose of the rights and property of whatsoever nature of the said Company nor cause the said Company to dispose freely of the same.
>
> Article #2—The Chinese Government agrees that all mines in the neighborhood of those owned by the Hanyeping Company shall not be permitted, without the consent of the said Company, to be worked by other persons outside of the said Company . . .[138]

Even though Yuan Shikai was humiliated by the Japanese and forced to accept most of the demands, he successfully refused to agree to the stipulations regarding the Hanyeping Company. Under the agreement, the Japanese did not take direct control over the Hanyeping Company, but they forced it to provide iron at below-market fixed prices and this added to the firm's financial difficulties.[139]

By the early 1920s the Hanyeping Company crumbled. Though iron continued to be processed at the Daye Iron Ore Mines using Pingxiang County coal, the ironworks in Hanyang stopped production in 1922 and never again achieved full production.[140] The Pingxiang County Coalmines, in fact, remained the largest Chinese-owned mines with a total investment of some $12 million, most coming from fixed-interest loans from Japan.[141] Reports in the West continued to support the notion that they were likely to succeed for the foreseeable future, putting the total holdings of Pingxiang County's mines at 300 million tons.[142] However, through the first half of the Republican period the level of output remained uneven and largely below the numbers hoped for by investors and economic planners.[143] By 1927, one report lists Hanyeping's debts to Japan alone as over ¥57 million against total assets of $52 million.[144] Finally, by the 1930s what was left of the Hanyeping firm was being financed by Japan and was essentially under Japanese control.[145]

As for Pingxiang County specifically, modern technology that was supposed to increase production for the empire greatly harmed the lifestyles of the local population. Those commoners who were hired by the Mine Bureau left the villages, their families, and their farmlands behind to work year-round in the mines. This strategy was supposed to allow the firm to spend little more than subsistence wages for labor and virtually none for the other family members. However, when the workers successfully protested for better conditions, Leinung decided to erect dormitories and hospitals to control them and provide them with their daily needs. Laborers in this way forced the Mine Bureau to invest in their basic welfare. Other commoners who took non-mining jobs in Anyuan were also affected by the changes in the economy. New markets and occupations allowed some to leave the villages to work as maids and servants, while others turned to running tea shops and opium dens and even to prostitution to provide services to the miners. Thus, instead of sending laborers to coalmines as a supplemental occupation that held families together, the new system began to pulverize the family into individual employees of the modern economy.

For local elites, while landholding and mine management had in the past been assets used to control labor and the inhabitants of the county under the new order, these powerful men were left to watch as outsiders outperformed them with capital and know-how beyond anything they could muster. Attempts to use their special access to the levers of government failed at nearly every turn. Finally, when the railroad was completed, the young elites looked forward to access to the outside world while their parents could only mourn the loss of the virtue of the highland lifestyle.

However, the Mine Bureau was also forced to negotiate with the local inhabitants and with itself as it implemented each of the features of mechanization. Sheng and Leinung constantly negotiated over questions of who could be hired for the task and how much it would cost. For the first two years, Leinung was almost certainly alone as the only foreigner in the county. Needing friendship as well as expertise, he repeatedly demanded that Sheng hire more German engineers to assist him with the project. And they eventually came. Sheng agreed to hire men only reluctantly and even then the numbers stayed low throughout the years. This meant that Leinung had to modernize the mines using local laborers who had no background in the technology. His poor labor force continued to hinder and cripple the plans he was making for the plant even as he kept envisioning a mine that would rival those back in his homeland, a vision that as it turned out never materialized. In fact, Leinung's plans of integrating modern technology and labor discipline into the Chinese community and his need for his modernization schemes to foster a society unified behind his goals led to just the opposite outcomes. As will be clear in the next chapter, modern technology created new avenues and topics of resistance while modernization led to new fissures throughout the society that could not be mended.

6

Social Atomization and Local Resistance

Divergent Desires and Strategies of Elites and Workers

While many studies have shown that urbanization transformed the worker population from peasants into a working class, many have also pointed out the pitfalls of this simplistic dichotomy. In his pioneering study, E. P. Thompson argues that the development of a working class in England was not a single event but that it was rather a gradual shift from a community of workers into a more solidified class-conscious body.[1] In some countries migrants never entirely broke their bonds with the countryside. Home villages provided familial ties and the prospect of a safety net when working conditions became too onerous to continue in the city. These ties allowed factory laborers to give up their status as working class rather than fight management over stipulations of the contractual relationship, and this, in turn, stifled the prospect of creating a full-fledged proletariat.[2] In these cases, the definition of factory laborer was constantly renegotiated as the ties to the home village and the urban factory tugged workers in both directions.

At the same time, the establishment of a factory did more than create the working class, as it also altered the lives of people tangentially connected to the new industrial plant. As workers contested their status by bargaining over pay, working conditions, and personal and collective status, the struggle in essence changed the nature of power among factory owners and managers. Workers who left their homes for jobs in the plants were forced to increasingly rely on the firm for daily subsistence, and that reality resulted in labor-management dialogs as both parties discussed issues like food and healthcare. These negotiations, in turn, had a significant effect on the villages and peasant family members whose role in the daily lives of the factory workers waned. For these people, workers ceased being part of the home. They were no longer family members who performed seasonal labor in a factory or mine and then spent the majority of their time in the family fields. And because of this, the physical requirements and world views of peasants and factory workers—even immediate family members—gradually changed and drifted apart, and this helped to sever their social and financial ties as well. The evolution of these changes moved through fits and starts, however, as communities often returned to those earlier relationships that were comforting and provided the best opportunities for success if not at least survival.

Moreover, social changes among the commoner classes, that is, workers and peasants alike, altered their relations with local elites as well. In larger cities, elites gave up their premodern roles as landlords and the holders of state-sanctioned virtue and turned to new strategies of philanthropy and new avenues that might lead them up the social and political ladder.[3] Following Marx's model of capitalist development, it could be assumed that these elites detached themselves from the patron-client relations they developed with the peasantry. However, this was not always the case. Rather, members of the newly emerging urban elite often maintained their ties with agriculture-based classes, allowing commoners to continue their control of the productive forces and their assurances of subsistence. Under these conditions the dialectic between capital and labor that Marx described that was essential for capitalist development did not happen. Rather, amalgamations of capital-intensive industrial production and labor-intensive subsistence agriculture were supported by semi-modern urban elites, premodern rural elites, and a collection of ad hoc institutions that to some degree supported them all.[4] And so a member of the working class who was subjected to a modern employer-employee relationship with his managers also maintained a clientage relationship with the landlord who owned the subsurface rights to the land his wife and children worked. At the same time, the member of the gentry elite who claimed the subsurface rights to large swaths of farmland in the county that he inherited from several generations past was being marginalized by a newly anointed elite who claimed superiority over him.[5]

Finally, even as the focus of historical change in Pingxiang County in the late nineteenth century was primarily due to altercations between and alterations of Chinese elites and nonelites, the impact of foreign engineers and imperialists provided yet another layer of friction and change. In these decades, much of the Southern Hemisphere came under the domination of European powers who sought to exploit their natural resources and labor power. While some of these empires were primarily occupations of hostile forces over an enslaved population, in other cases, such as in China, the nature of Western domination varied from location to location and evolved over time.[6] In the case of the Pingxiang County Coalmines, specifically, Western influence was mostly experienced through personal interaction with the German engineer Gustav Leinung and his associates. Since these foreign experts arrived with technological knowledge but without military force, they were required to use the power of persuasion to make the changes they believed were necessary for technological success and economic growth. At times their needs were met with the support of Chinese officials, local elites, or workers, and at times their needs—like the desires of each of these sectors in the Chinese of the county—were not met. As each sector sought to find a means of attaining their personal aspirations, they established various coalitions and alliances and fought against other coalitions and individuals. Eventually, the Germans struggled against the Chinese, and Chinese and German elites fought against Chinese peasants and workers. Over time, under the pressures of technological and social change, the community became more atomized and broken and this sparked greater levels of desperation and violence.

To articulate the new parameters of Pingxiang County's society, I focus in this chapter on a small number of both minor and significant examples of negotiations and struggle that transpired during the last years of the nineteenth and first decades of the twentieth centuries. These events point out a desire among the county's members for contradictory goals of change and continuity. Workers fought to be treated as professional laborers in the mines while clinging to their agrarian ties with the countryside. Local elites integrated modern thinking and technology with their own emerging sense of nationalism, while at the same time they opposed the foreigners and outsiders who provided the tools that made this possible. Moreover, even as elites were thinking of a new time when the Qing dynasty would not rule the empire, they led the militia against anti-Qing revolts following the precedent of the Taiping Rebellion five decades earlier. For outsiders like Mine Bureau leader Zhang Zanchen, these men emerged from obscurity as petty merchants to become modern industrial managers. However, while they thrived on their *new* status as part of the forward-looking economy, they relied on their *imperial* status for their authority over laborers and local elites alike.[7] Finally, even as Gustav Leinung described Chinese labor in demeaning tones and ruled over the workers with efficiency and brutality, he also found a sense of beauty in Chinese culture that took him beyond the world of steel and stone that he was hired to create. For each of these people and the strata they represent, modernization altered their needs and desires, their view of their world and themselves, and the strategies available to them to achieve their goals. Therefore, in this chapter, I show that, unlike Marx's depiction of a gradual but inevitable transformation from some precapitalist world to a capitalist one, that in Pingxiang County at least, the changes were complex and omnidirectional, they moved in fits and starts, and even as they moved toward a new order they at times averted back to the beginning.

Opposing the Greedy Wolf: Elite Resistance

Because elites and nonelites experienced the technological changes in the coalmines differently, they held different views regarding the promise this development held for their own futures. As will be discussed later, workers and peasants viewed the mines as possible avenues of income, and therefore they fought over working conditions and wages. On the other hand, lineage leaders who were educated and land rich saw the coming of the new mining scheme as undercutting their privilege. While at first the gentry—particularly the Wen lineage—supported the official policies of Zhang Zhidong, after the state forced them to sell their mines and give up their incomes from those rents, they felt betrayed by the officials who had occupied the town of Anyuan. Furthermore, while the Confucian-educated gentry used their connections with government officials to maintain their status and uphold the principles of the state, their modern-educated sons viewed the coming of foreigners with dread and sought to hinder if not halt the mechanization process that these same officials planned for their county.

For at least some members of the elite, opposition to Western influence began even before the German engineer, Gustav Leinung, arrived in the county. In the late summer of 1896, newspapers began to publish information of the impending arrival of Leinung. One such paper, the *Han Bao*, wrote in part that Leinung was coming due not to the wishes of Sheng Xuanhuai but of the local gentry leaders when it indicated that "a foreign mine specialist was invited by gentry Wen (Tingshi) to develop the coalmines."[8] This story of a single foreigner arriving in their highlands might have seemed innocuous enough. But, to the young students of the Jiangxi provincial highlands who were aware of the violent confrontations between foreigners and China taking place along the coastal areas and port cities, the implications were frightening. By the early fall, as Leinung ascended the mountains traveling in from Yichun County into western Pingxiang County, students were in both county seats to take the county-level exams. Realizing that his arrival was imminent, they started to contemplate some form of action. To this end, the students in Pingxiang County openly spoke out.[9] In September a series of big-character wall posters—a means of protest used by young Chinese scholars for centuries—appeared on the doors of several Confucian and education halls in the city, accusing Wen Tingshi of conspiring to open Pingxiang County's natural resources to imperialist greed.[10]

Though mostly unsigned, the posters were almost certainly the work of students taking the civil service exams and who were also sons of the county's elites because they were highly literate and were hung in the gentry-dominated district of the county seat.[11] One anonymous letter posted compared the intervention of foreigners into the Pingxiang County mines with past flooding experienced by the inhabitants. The author also indirectly accused Wen Tingshi when he wrote, "Recently, I heard somebody in Pingxiang [County] enticed some foreigners to exploit coalmines here and made a ten-year contract with them without permission. Won't they make another contract ten years later?"[12] The author predicted that the foreigners would destroy the fields and gravesites and harm the *fengshui* contained in the mountains. He also envisaged that the foreigners would want to exploit the silver mines and the iron mines and that they would use modern machinery to replace laborers and therefore take jobs away from the people. Finally, the author claimed that foreigners would mistreat the Chinese inhabitants, and coerce people to join Catholic churches and thus undermine the traditions of the community.[13]

Another letter, signed "Government Students," was more direct in its accusations. Based on the *Han Bao* article it said in part:

> Concerning a Jiangxi City, our newspaper reports that "a foreign engineer will arrive and stay in the city god's temple." We are told that the engineer has been employed by [Zhang Zhidong]. Scholar Wen Tingshi persuaded [Zhang] that his hometown has coal of fine quality that can be used by machines and steamers and that he wants to pull resources together to open the mines and increase their benefit. [Zhang] has agreed and has commissioned Wen to go with the foreign engineer to check the quality of the coal.[14]

The authors of the letter used the *Han Bao* article to accuse Wen Tingshi of inviting the foreigners to come to Pingxiang County and, working in collusion with the Western engineers, to plan to engage in profitable actions at the expense of the local population. The signers called for acts of violence against the foreign intruder and put the elites who supported modernization on notice that such actions would lead to harsh reprisals.[15]

In another anonymous article, one of the longest, the authors set out many of the same points found in all the earlier letters. The authors ended with an interesting metaphor when they argued,

> To attract foreigners to Pingxiang is like opening your door to invite robbers. When they see the treasures in your home, they will take them all. When they see the treasures of our county, they will exploit them to the utmost. The possessions of one house and one county are limited. After a hundred years of accumulations, they are still not enough. Yet it takes only an instant to squander them. . . . [The foreigner's] nature is that of a wolf, greedy to no end. They should be fought against and driven out, not attracted [but] poisoned.[16]

Taken together, the posters indicate that the students were aware of the implications of foreign intervention in their community. While they were benefiting from the expansion of modern newspapers and discourse, as evidenced by their comments, they were at the same time fearful of the changes industrialization could bring and even suggested some of the xenophobia that was stereotypically said to be part of the region's culture.

These letters, and perhaps others that were not subsequently transcribed, were confiscated by Pingxiang's county magistrate, Gu Jiaxiang. Gu was from the east coast of China and had more experience with Westernization—for better and for worse—and as the county's magistrate was charged with the responsibility of both doing the government's work and maintaining the peace. Having read the letters and studied the information pertaining to their posting, he realized that the students were directly criticizing the court's wishes, and so he strongly criticized them for their intransigence. He wrote a rebuttal that both criticized the content of the posters and warned students of repercussions if they continued their actions. He argued that the author of the *Han Bao* article did not directly implicate Wen Tingshi in the acts of collusion put forth by the authors of the posters. Also, Gu understood that court officials sent Gustav Leinung to investigate the county's coalfields. Hiring Leinung "is just like inviting a doctor," he wrote, using a very different metaphor than the student letter writers, "What he can do is prescribe. As to how much or how soon to take the medicine, the doctor has the final say."[17] Then, in a separate letter, Magistrate Gu discussed several points raised by the posters specifically. The German engineer was sent to prospect the local mineral deposits, he explained, not to mine them, and thus it was not certain that the foreigner was going to extract any of the coal. Moreover, Zhang Zhidong and his assistants were not going to consider the silver, gold, and other precious metals found in Pingxiang County. Gu also explained that the county's *fengshui* would not be destroyed and the Catholic Church was not going to come to Pingxiang County. In a subsequent letter the magistrate explained that foreign machinery would not be needed and therefore there should be no disruption to Pingxiang County lifestyles. He argued in part that, since China had a much higher population density than Europe, modern machinery would not be required. The high cost of machines, he explained, would not be cost-effective where cheap labor was so plentiful. Thus, the argument that Western coal extraction practices were about to remove local miners from their jobs was unfounded.[18]

Gu also used his first letter to harshly criticize the students from the countryside whom he believed authored the posters. He apparently assumed at the time that the posters were written by students who were usually with their families celebrating the Moon festival in the countryside at this time of the year. In fact, he felt particularly constrained by the events because there were no urban gentry in the city to help him with these matters. Furthermore, several gentry sent Magistrate Gu a letter opposing the foreigner's arrival, a letter that is not extant but that might have provided more evidence of the antagonism of the countryside's elites.[19] In any case, after he laid his claim at their feet, Gu criticized the gentry families from the countryside, telling them they must exert better control over their children when they arrived in the city. He threatened to take government stipends away from students who behaved in this manner in the future.[20]

Because Gu Jiaxiang felt the students were likely to incite violence against the arriving foreigner, he ended the exams early and sent them all home.[21] He also advised the traveling contingent bringing Gustav Leinung to remain outside of Pingxiang City until after the exams ended. Leinung and his Chinese companions passed quickly through Yichun and quietly proceeded to Luqi in western Pingxiang County, where local militia protected the foreigner.[22]

Even though the exams ended early, some of the students stayed in Pingxiang City and witnessed Leinung's arrival. When the delegation entered the city, they lodged in Shangbin Hall, a building located in the gentry district. At least one letter of opposition was posted on the door of this hall, its author claiming to be appalled that a foreigner would be housed in such an important building. However, Gu explained to the hall manager in a subsequent letter that there was no reason to oppose Leinung. Also, Gu noted that the Guest House in Shangbin Hall was the only location in Pingxiang City available at that time to house the entire delegation.[23] The students in the city were aware that Leinung was housed in this building, and several of them displayed their anger toward the foreigner by trying to beat the manager of the hall. Magistrate Gu, who arranged for the guests to stay in the Guest House, scolded the students and quieted the mob. Not only did Gu use his ability to write letters in support of the modernization, but he also was an effective speaker, a skill he often employed to subdue the local elites. Hui Jixun, one of Sheng Xuanhuai's appointees, explained that magistrate Gu "had been in office for years and could convince people. However, after talking his mouth dry, the people were only a little persuaded."[24]

After Leinung arrived safely in Pingxiang City in October 1896, the county government provided military protection and by various means sought to accustom the people to the foreigner's presence and calm the simmering unrest among them. In one description based on Leinung's recollections of his early weeks in the county, one author wrote:

> Upon his first arrival in [Pingxiang County] he was an object of awe and amazement to the natives who had never seen a European. All sorts of beliefs were held about him. He was said to have three eyes, one at the back of his head, and it was believed that he could see deep into the bowels of the earth and detect the treasures there. When he entered [Pingxiang City] for the first time in 1896 the natives were sitting on the roofs of houses to see him. He was put into a small room in an ancestral hall in which there was a grated window, and the people were allowed to come along in squads of 10 or 12 to have a look at him, as though he were a rare zoological specimen. The idea was to make them accustomed to the appearance of a foreigner. During the first year there were always 200 soldiers around Mr. Leinung. He was not allowed to leave his quarters and had strictly to obey orders.[25]

This depiction of the early months provides evidence of a very tense period throughout the county. Pingxiang County's local militia forces in the early twentieth century numbered about 250 men, so a contingent of 200 troops as described above was very large indeed.[26] Once his appearance no longer shocked the city residents, he was moved to the Guangtaifu Lineage Trust headquarters, located in the newly constructed Wen Lineage Temple built by the Mine Bureau. But even here, putting him up in this structure located just outside the city walls and miles from the coalfields, suggests that the county government and gentry supporters feared for his safety.

Yet even as Leinung was accepted into local society, the machinery and technology he infused into local mining was similarly a focus of anxiety. Once the German engineer's modern machinery arrived in the county, the local population viewed these loud metallic contraptions as both an unknowable burden placed on the people as well as a tangible version of the evil that awaited them. The local scholar Zhang Guotao writes that the people feared what they did not understand and worried that their world was being destroyed in the interests of foreign insatiable demands:

> The invasion of backward areas by modern enterprises and imported goods often evokes general resistance from the established forces. So it happened in [Pingxiang]. A weird variety of rumors circulated through the county seat and countryside. Some contended that the railroad destroyed [*fengshui*] . . . , thereby disturbing the ancestral graves. Others insisted that a child had to be fed into the locomotive's smokestack each day before it would run, and the same was reportedly true of the chimney at the coal mine. There was deep-seated hatred for such new monsters as trains and mechanized coal mines.[27]

Zhang's explanation of his community's trepidation is interesting for several reasons. First Zhang explained that the fear he described was common throughout Pingxiang society. The ideas regarding the horrors of the smokestacks were not simply limited to the peasant population but instead "circulated through the county seat and the countryside." Also, as was discussed in chapter 4, the people feared the railroad and other machinery could disrupt the county's *fengshui* by altering the topography and significant religious sites. More telling, I would argue, are the stories and descriptions that personified machinery using religious or supernatural imagery that indicated that the new technology was not simply anxiety provoking but was evil in their minds.[28]

Even as students used their scholarly status to try to embarrass the local government, more senior gentry pushed back as well. The gentry leader Xiao Liyan became a harsh critic of the industrialization project in Pingxiang County and he worked in various ways to extract wealth from the project, if he could not undermine it altogether. Xiao received his *jinshi* degree in 1892, was singled out by the emperor as a particularly gifted student at the exams in Beijing in 1894, and was awarded a position at the prestigious Hanlin Academy in Beijing. Two years after taking that post he was back in Pingxiang County. And in February 1897, Xiao Liyan and twenty other gentry leaders signed a petition to Sheng complaining of the initial intrusions of Lu Hongchang and his associates that is discussed fully in chapter 3. The signatories are nearly all ranked in order of their degrees with the top-level *jinshi* holders signing first and the second-ranked *juren* recipients second. However, even though Xiao was one of the newest holders of the *jinshi* degree, his signature is first, indicating that he may have had more to do with the petition than some of the others.[29]

Similarly, in a subsequent letter dated 1899, Xiao and several other gentry members—his name appeared second—criticized Sheng Xuanhuai's plans to secure large loans from Germany in order to purchase a railroad. First, the signatories charged that the building of the railroad from their county to the Xiang River valley was not a good plan for their county. They specifically argued that the Lu River was suitable for all the needs of the ironworks, though they admitted that some seasons were better for boat traffic than others. These gentry leaders further explained that since their county was mountainous, the railroad's route would be circuitous and would require many bridges and other difficult passages along the way. Furthermore, the expansion of mining had also already led to more jobs for commoners, and the creation of a railroad would jeopardize those jobs and their livelihoods. They asserted that this plan was not devised in the interests of the people who would be impacted the most.

The second argument in their petition was that the cost of the planned modernization projects as a whole threatened to put the empire into debt that China could not escape. They began by stating that they had heard that the initial loans to Sheng Xuanhuai from Germany would total some 4 million marks, which they argued was a bad investment given the circumstances. The coalfields of Pingxiang County, they explained, were fine for local consumption but would not provide the wealth required to pay back such huge sums of money to the German banks. Thus, if this plan went through, the empire would drown in debt to foreigners.[30]

Xiao's antagonism became a sore spot for the mine's leadership as seen in a letter written November 11, 1900, by Mine Bureau manager

Zhang Zanchen, that accuses Xiao of organizing gentry leaders against the industrial project in the county. Zhang began his rather emotional letter by complaining that "Jiangxi is old-fashioned" and that "officials and local people alike are against the railway and the Westernization Movement" (*Yan wu yung dong*). He explained that Gu Jiaxiang was doing a fine job purchasing lands to complete the railroad but that Xiao Liyan was organizing other gentry against Gu to try to harm him and thus stop the railroad from being completed. Xiao and his supporters, Zhang wrote, believed that the railroad was harmful for the county and that it would cripple the local economy. And he went on to state that even though Xiao had been awarded a position as a Zhou magistrate in Shanxi Province, he threatened to turn down the position so he could stay in Pingxiang County and continue to work against County Magistrate Gu. For the immediate period, Zhang Zanchen called on Gu to leave the county seat and continue to purchase lands in eastern Pingxiang County. However, Gu wished to step down from the position and take a government job elsewhere. Specifically, he hoped to leave the county behind for fear that Xiao would attempt to have Gu brought up on charges of some sort.[31]

Zhang did not want Gu to accede to the whims of Xiao Liyan and his supporters because he felt that only Gu could do an adequate job protecting the railroad. He thought that if another magistrate was brought into the county at such an important time, Xiao and others would sense that the government was weak and could be manipulated. For these reasons, Zhang requested that Gu be kept on as county magistrate until the railroad was completed.[32] Zhang's superiors granted this request as Gu continued to serve as the county magistrate until the line was completed to Liling City less than thirty miles from the terminus at the Xiang River in 1902.[33] Also, apparently staying true to his words, Xiao Liyan remained in Pingxiang County until Magistrate Gu left office. Finally, in 1902, Xiao accepted a post as a Zhou magistrate, where he served with distinction.[34]

Subsequently, after Xiao retired from public office in 1905, he returned home to help modernize a local school. When he arrived in his home county, he argued that the costs of improving the school—which included refurbishing the building, hiring a teacher, and locating students—could be paid for by some of the profits of the Pingxiang County Coalmines. He calculated that the academy needed 2,000 piculs, or more than 260,000 pounds, of rice yearly to fulfill its needs.[35] To this end, he again organized members of the local community to extract money from the coffers of the Mine Bureau. And this time, with a new county magistrate, Xiao was able to get an agreement from the county yamen to assist him in this endeavor and bring the Coalmines to award these funds to Xiao for the purpose of modernizing the school.[36]

While Xiao Liyan was using his status as a scholar and official to block or hinder the mines, Li Youru, the mountain lord and member of the powerful Li lineage, fought to take back his mines and those of his allies. Joining together with other like-mined elites in an organization called the Sustain the Mine Society, Li argued that the Mine Bureau had taken all the good mineral fields, including, presumably, his own. These men further claimed that the lands purchased by the Mine Bureau expanded far beyond the boundaries agreed upon by the court based on the memorials sent by Zhang Zhidong and Sheng Xuanhuai, the two founders of the modern mines. They charged that the Mine Bureau's insatiable desire for the county's properties was ruining the lives of the people of the county and that the monopoly on the county's minerals meant that "more than tens of thousands of people were left outside facing the [mine's] perimeters and sighing."[37]

Similarly, Wen Tingshi and his allies, the leaders of the Guangtaifu Lineage Trust, and the men most closely connected to the initial actions of mine mechanization were attempting to hinder expansion of the Mine Bureau's ownership over an important iron ore mine located in their township. The conflicts over iron ore fields in the Shangzhu Mountains pitted local concerns against several government and industry forces. These iron ore fields were located near the county's western town of Xiangdong near Guisheng Township, the chief location of the Wen lineage and Wen Tingshi's extended family. They contained the finest quality iron ore in the county. And in the first half of the nineteenth century some excavation had been completed to systematically extract mineral from the site. However, initial attempts failed due to inadequate technology and insufficient demand for the product. Then, by sometime in the late nineteenth century, the lucrative deposits and favorable location led one Hunanese provincial governor to attempt to take control of the mountains presumably to add to the Xiang River valley mineral markets discussed in chapter 1. This plan sent the local gentry to action, fearing the loss of what they viewed as their community's resources.[38]

Though the timing of this conflict between the Hunanese provincial governor and Pingxiang County gentry is not clear, it may well have inspired the Wen lineage to attempt to make an agreement with the Pingxiang County Mine Bureau regarding the iron ore mines. In 1896, while the Wen lineage leaders were establishing a working relationship with the Mine Bureau through their Guangtaifu Lineage Trust, Wen Tingshi formally requested the state to engage in iron ore mining in the Shangzhu Mountain range. Since iron ore deposits were especially important in the state's drive for industrialization in the last years of the nineteenth century, his proposal whetted the appetites of the state and the leadership in Hanyang. In 1897, Zhang Zhidong had the mountain's iron ore tested by a Belgian metallurgist who concluded that the deposits in Pingxiang County were perhaps superior to those found at Daye, the principle iron ore mine for the Hanyang Ironworks. Sheng Xuanhuai subsequently received permission from the imperial court to develop the mines. The court sent a delegation to the county that included engineers from Belgium, Germany, and England who determined that the iron ore mines were suitable for the needs of Hanyang's industrialization project. Therefore they called for a second mission to begin the management of the iron ore mines, and those men were soon sent.[39]

By this time, the Mine Bureau forced the Wen lineage leaders to sell them their mine properties, and this all but ended any chance of collaboration in future endeavors. In 1898, soon after losing the lineage's firm, Wen Tingshi was implicated in the court-centered Hundred Days' Reform and he briefly fled to Japan in fear for his life.[40] And while he was away from Pingxiang County, Leinung and local gentry began sparring over the issue of mining iron ore in the Shangzhu Mountain range. In an entry in the 1935 gazetteer, Duan Xin, a fellow gentry member and resident of Guisheng Township, explains that Leinung decided to mechanize the iron ore mines, and this led to a new series of conflicts with the Wens and others. Having previously investigated the iron ore mines himself, Leinung determined that the quality of the minerals was suitable for the manufacturing of tools and machinery needed on the factory grounds. To this end, in December of 1901, he brought the mines under the control of the Mine Bureau.[41] Then in mid-June of 1903, around the time Wen Tingshi returned to the county, Leinung wrote a memo to Sheng Xuanhuai advocating the development of an ironworks in Pingxiang County utilizing the iron ore in Shangzhu Mountain. He explained to Sheng that the iron ore from the Shangzhu mines would be more expensive to extract than the mineral in the Daye Iron Ore Mines but that the cheap cost of locally produced coke would keep the prices comparable. He did, however, concede that the price of constructing another branch railroad line to the Shangzhu Mountain range could cost more than 100,000 taels, a price that he felt was nearly prohibitive.[42]

While Leinung and Sheng were trying to work out the expansion of industrial development in Pingxiang County, the local gentry utilized their status and collective capital to try to hold off these men's desires. In 1903, as Leinung's memo about his designs for a small ironworks in the county circulated, Wen Tingshi and several fellow Guisheng Township gentry members joined together to buy up their native township lands containing the iron ore mines on Shangzhu Mountain. Wen and his allies hoped that the iron mines could be handed over to the local county government and thereby protected from outside exploitation, a plan that apparently failed.[43] These actions point to a change of heart by Wen Tingshi, who was transformed from a supporter of modernization under the Guangtaifu Lineage Trust to an opponent of Mine Bureau actions in his homeland. Wen Tingshi's personal actions regarding the mines did not come to full fruition, however, because he died in 1904 before he could implement his plans. However, several of his allies did manage to buy the iron ore mines in the early Republican period while the political unrest hindered the government-run Mine Bureau from continuing its work. With the mines under their ownership, they closed them to outside influence and development and restricted shipments to other counties and markets while providing iron ore to the local population.[44]

Duan Xin, in his gazetteer entry about this incident, conceded that local control over the mines was likely to be temporary as the demands by industrialists continued to place pressure on the county. However, he argued that the project as it stood at the time of his writing sometime in the first two decades of the twentieth century was superior to the "rut" the local population experienced with the mechanization of the coalmines.[45] It does appear, however, that the success was temporary because the iron ore mines were once again under the control of the Mine Bureau by the mid-1910s and had been secured even against the wishes of the most powerful lineage member in the county.[46]

Wen Tingshi's argument with the Pingxiang County Coalmines was very different from the students' complaints. While Wen simply wanted to hold on to his community's local resources, the younger generation viewed the coming of mechanization as part of China's global struggle to hold on to not only its resources but its way of life, values, and beliefs. Yet, in each case, these men assumed that they had the right and authority to push back against outsiders using their elite status and the gentlemanly weapons available to them. They wrote big-character wall posters and petitioned their superiors to coerce the state to accede to their needs. In the rest of this chapter, on the other hand, I show that when the mining operations altered the social economy in ways that encouraged the commoner classes to engage in violent unrest, some of those same elites returned to their former mind-set as the holders of Confucian order. They put away their differences with the officials of the Mine Bureau and inflicted ruthless force against members of their own county and lineages.

"Local Hooligans": Commoner Negotiations and Unrest

While the elites negotiated with Mine Bureau officials to try to maintain their social status and their control over minerals and labor, the level of brutality traumatized the miners, who feared that the new scheme would end their families' subsistence needs in the villages. The conflicts of commoners against the mines began when the news had reached the county that a foreigner was coming to test the local coalfields. In the late summer of 1896, upon hearing of the imminent arrival of Gustavus Leinung, the miners believed that the Wen lineage had betrayed them by using the Guangtaifu Lineage Trust to take control of smaller mines, eventually forcing them to work for foreign leaders, perhaps under poorer working conditions than they had before. And so to show their indignation, miners initially engaged in petty acts of sabotage and violence. When Wen Tingjun and Xu Yinhui arrived at the mines in the summer of 1896, they complained to officials at the Hanyang Ironworks that miners were protesting attempts to intensify production. They wrote:

> Unfortunately there are some local hooligans who in the night released water from the mountain ravines so that it flooded and damaged the coking ovens. Inside, the ovens were also flooded. So we have sent a notice to the Pingxiang County [government] to send three police to arrest these people who released the water.[47]

Sabotage of this type suggests that the level of antagonism had not yet led to general unrest. Those "local hooligans" who flooded the kilns were probably a small contingent. They therefore had to use methods of stealthy destruction to hide their numbers but still make their point. However, acts of vandalism and encroachment by local miners on the holdings of the Mine Bureau continued for many years, indicating that the antagonisms that were expressed in these actions were not alleviated.[48]

Resistance became even more pronounced when Gustav Leinung entered the county. When he was traveling from the east through Jiangxi Province into Pingxiang County, during his initial trip to excavate the mines, local miners planned to attack him with rocks and other crude weapons. After the peasants and miners learned that the foreigner was going to be traveling into Pingxiang City from Luqi to prospect the mines, they tried to stop him. They gathered above the route between the two cities and sought to ambush him with a hail of rocks when he passed by. Fearing the future under the Western engineer, they hoped to stop the encroachment of modern and foreign schemes from overtaking their mines and thus their livelihood. Fortunately for Leinung, his contingent found out about the plot and took another route to Pingxiang County that allowed him to arrive safely.[49]

While these actions were pitifully feeble, as mechanized mining came to Pingxiang County the inhabitants gradually changed and evolved as they became better organized and angrier with the lifestyles the Mine Bureau and others imposed on them. Even as the gentry leaders sought to hinder excavation or profit from the mines, the Gelaohui secret society began actively organizing Pingxiang County as well as Liuyang and Liling counties in Hunan Province. It began as a semi-religious and semi-political movement that sought to overthrown the dynasty, and it gradually evolved into a supporter of commoner desires for a complete overturning of the sociopolitical order. This organization sprouted up through various townships and districts and grew until the majority of the peasant population made up its membership. In Anyuan, as the miners concluded that simply throwing stones at foreign managers was insufficient, the Gelaohui organized virtually the entire workforce through the contract labor system. Its organization became so powerful in the mines that it established its own headquarters in the town and hand-selected the mine bosses responsible for the workers in the mechanized mines.[50] The umbrella organization of the regional secret society included a coalition of ex-soldiers, charismatic workers and peasants, and urban intellectual nationalists. And these society chiefs organized peasants and workers through informal and illicit channels that undermined if not replaced the formal institutions of the imperial state. It also replaced significant aspects of the patronage systems that dominated the lives and economies of the local population as well. To this end, the Gelaohui leaders not only insinuated themselves into the positions of the mine workers' labor bosses but also became a protection system against landlords and government officials for the villagers in the countryside. This alliance signaled a growing level of animosity of people being crushed by the failing empire, the rising West, and the lack of adequate resolutions to their collective suffering.

By the end of the first decade of mechanized mining in Pingxiang County, tensions increased to the point where the coalminers began to contemplate violence. The fact that Leinung had recruited most of his miners from Pingxiang and Liuyang counties, two Gelaohui strongholds, exacerbated those tensions. Other workers he hired, including those who worked on the railroad, came from Liling County, another secret society stronghold.[51] Under the stress of longer hours, more accidents, and continued poor wages, secret society–backed miners, who likely numbered at least a thousand or so workers, began to confront their German managers, who by this time numbered no more than five or six.

The first open confrontation happened in 1904 when the German engineers reduced the pay of the workers, charging them with diluting their coal hauls with shale. At the time, Leinung was in Europe and so he temporarily placed another German engineer in charge of the mines.[52] When the workers found out that the managers were reducing their pay, they went to the German engineer in charge. The miners argued that they were not responsible for the problem, but the German engineer demanded that the workers take cuts in pay or improve the quality of the mineral. The negotiations were tense and probably led to some concessions by the German engineer. Subsequently, when Leinung returned, he was angered to find that his assistant bungled these confrontations. He immediately fired his subordinate and renegotiated the contracts regarding the quality of the coal to be more in line with his own strict standards.[53]

Almost as soon as the conflict was resolved, Leinung built dormitories and cafeterias that he hoped would break the miners' need for the contract labor bosses. Understandably, this attempt at changing the relationship between the miners and the mine angered the contract labor bosses and their secret society leadership. In an attempt to try to stop Leinung's actions, the miners and their bosses met to devise a strategy against him. During an investigation after the fact

> [i]t leaked out that on the fifth day of the New Year the miners had been assembled at a temple nearby and that one of the gang leaders had addressed them telling them that the proposed innovation was against their interests and that they should combine to protest against it. The usual thing happened; the crowd went with the agitators and the aid of a *literati* had been invoked to draw up the protest. They all swore not to betray each other.[54]

After agreeing not to "betray" one another, the miners briefly struck against the dormitories and cafeterias, arguing that they were not willing to work under Leinung's conditions. When the miners lost their resolve, and went back on their oaths, the workers' collective action quickly broke down. Leinung fired the contract labor bosses he did not trust and told the miners to go back to work.[55]

However, this did not end the problems as more issues erupted, increasing antagonism between the two factions. In 1905, a gas explosion inside one of the mine shafts killed a reported ninety miners. As other Chinese miners ran toward the explosion hoping to save some of the workers, German managers chased the men away for fear that more would be killed and closed the shaft, leaving any injured miners to die inside the mountain. When the men put out the fire and the danger of the accident was over, the German managers ordered miners to reenter the mine and collect the bodies. After the dead were removed, mining resumed in what many miners regarded as a callous disregard for the victims of this tragedy. In fact, the workers viewed the incident with such horror that they named the shaft "The Bake People Oven."[56]

Incidents such as these struggles over pay and safety heightened the tensions between the workers and German managers and increased the possibility of violent confrontations. Leaders of the Gelaohui secret society—particularly the well-known leader Ma Fuyi—further organized and incited the workers. Ma was a peasant from Hunan Province who briefly fought beside the Nationalist Huang Xing in 1904 in one of Huang's abortive peasant uprisings that sought to overthrow the Hunanese provincial government. Now, in 1905, Ma turned to the miners at Anyuan to organize another uprising. He told the workers of his anti-Manchu beliefs and his plans for a revolutionary movement. Because the miners were already in conflict with the German managers, the attempt to spark violence in the mines was almost automatic. Moreover, even as he was gearing the miners up for rebellion, Ma established himself as a hero to the villagers in the highlands and organized them to fight with the miners in a general uprising. However, as Ma's plans to lead a rebellion of miners and peasant supporters were about to come to fruition, he was arrested and executed for attempting to overthrow the government. Peasants and miners alike viewed the execution as a blow to their cause and a savage act on the part of the government.[57]

At about this time in May 1905, Leinung went to Shanghai and placed another engineer in charge of the mines. Once again the miners tried to force the Germans to accept a reduction in the quality of the coal they extracted and also sought an increase in pay. In addition, they demanded the reduction of their day from a twelve-hour to an eight-hour day and the creation of a third full rotation of workers. The new German manager, eager to appear strong-willed in the face of the firing of his predecessor, refused to give in to the Chinese demands and stopped their pay. When the miners found out that they would receive no pay, they sent a delegation to the manager's house in the German compound. Once they arrived, the manager again stated his refusal to pay them. The news sparked anger among the miners, who turned violent and began throwing rocks at his house and the other houses in the German compound. By the time the miners finished venting their anger, much of the foreign compound was destroyed, as were several administrative buildings. The manager became frightened and fled into his house.[58]

Not only was Leinung in Shanghai, but also the Chinese manager, Zhang Zanchen, was away taking care of other matters. In his absence, Zhang's assistant took the responsibility of protecting the German engineers, who were becoming increasingly concerned for their lives. He arranged for the foreigners to quickly board a train and flee into Liling City, which at that time was the terminus of the railroad line. When the news reached Zhang of the unrest, he quickly returned to Anyuan, pacified the miners, and allowed the Germans to return. However, even then the German engineers did not dare enter the mines for at least several days until Chinese officials successfully imposed their will over the angry laborers.[59]

The unrest of 1904 to 1905 existed almost entirely between the German managers and the Chinese laborers. However, since the Gelaohui was beginning to involve itself with miners and peasants alike, unrest in one sector of the economy or one location of the Jiangxi and Hunanese provincial highlands sent waves of political activity throughout the region. Given the dramatic changes the commoners faced, including not just the mechanization of the coalmines but the greater political demise of the Qing state, the region experienced resistance and rebellion in many forms throughout much of the late nineteenth and early twentieth centuries. Since peasants and workers struggled against landlords and external powers even greater than the lineage leaders who ruled their communities for generations, the Gelaohui secret society and its coalitions of gamblers, smugglers, and local toughs acted as their newfound protectors and patrons of many members of the community. And it was this coalition of forces—along with the assistance of young nationalists who were loosely connected to Nationalist movements led by Sun Yatsen, Huang Xing, and others—that came together to the greatest degree during the unrest in the following year.

A series of natural disasters and the subsequent impoverishment of the rural population sparked general unrest in 1906. Heavy rains—the worst in two hundred years—began in October 1905 and did not let up until June 1906.[60] Massive flooding in the spring along the Hunanese provincial lowlands reached the roofs of the houses and corpses floated down the Xiang River. The British consul, Bertram Giles, hindered attempts by the Hunanese governor to stop shipments of locally grown rice from being delivered to urban centers in other parts of the empire. Giles pointed out that treaties signed by the Chinese and British stipulated that rice and grain shipments could not be stopped for any reason without twenty-one days' notice.[61] Under this agreement, the Hunanese governor was forced to ship grain out of China's "rice bowl" while his own people starved. Subsequently, in the fall of 1906, a severe drought destroyed much of what was left of the agrarian countryside in Liling and Liuyang counties.[62]

The effects of these two disasters greatly harmed the towns and cities and the greater countryside of Pingxiang, Liuyang, and Liling counties as well. In time of such famines, it was often the case that villagers fled to the cities in hopes of taking advantage of the ever-normal granaries located in the county seats.[63] And in 1906 when the county's villagers arrived in Pingxiang City to beg for government help, they found themselves joined by thousands of other victims who arrived from the Hunanese counties because the granary systems in their counties were depleted or in disarray.[64] In particular, railroad workers and their families living in Zhuzhou, the railroad's terminus in 1905, bore the brunt of the famine. After the disasters ruined their homes and devastated their families, some of them wandered to Pingxiang County assuming the Mine Bureau would provide them with the assistance they required. Accounts of the famine tell of starved and stricken peasants dying along the roads in the Jiangxi highlands.[65]

However, when the people arrived in Pingxiang County, they did not find the assistance they desperately needed. The gazetteer contains two entries describing a horrific famine by a Pingxiang County scholar named Wu Shizhang, who was a tribute student in 1893, thirteen years before the famine.[66] Wu's descriptions, titled "Starving Masses" and "Expensive Grain," are not dated. However, one of them refers to a period of successive flooding followed by drought that fits the events of late 1905 and much of 1906. He wrote that the disasters brought thousands of people into Pingxiang City begging for food and money. The poor came in little more than straw sandals, walking "like fools, as if they were in a drunken stupor." They sold or pawned their clothes and some families even contemplated selling their children. Wu was told that over nine thousand people assembled in the city begging for food. These were good people, he reasoned, who now stood naked in the streets.[67] And yet their cries were not met with acts of kindness by the elites of the city. Instead, the rich families used whips and bamboo rods to keep the poor from overrunning the granaries. In fact, due to the corruption of local officials and elites, the granaries in many of the villages had been allowed to go empty by this time. The landed elites were unwilling to assist in the emergency needs of the local population.[68] Specifically, Wu wrote that in the towns and cities powerful merchants, or "street headmen," manipulated the price of grain through merchant alliances to make even more money in the marketplace, acts that no doubt further impoverished the people in the towns and villages. These acts, he argued, were not based on the basic precepts of Chinese society. Rather they were acts of "tricky local traders and villainous merchants" together with officials who, he stated in a hyperbolic tone, "worked to take the commoners' flesh."[69]

Once the flooding abated in the spring of 1906, many of the affected peasants from Liling County left Pingxiang City and returned to their homes hoping to restore their fields for the fall harvest. In an attempt to further assist the peasants, the Zhuzhou Railroad Bureau chief wrote a letter to Zhang Zanchen at the Pingxiang County Coalmines requesting financial assistance for the railroad workers and their peasant families. In an incident that once again struck the workers as uncaring and insensitive, Zhang wrote back that the financial situation of the mines at that time was weak and that he could not assist the railroad firm and its workers.[70]

This inaction of the Mine Bureau toward its workers and the continued indifference to accidents and working conditions sparked a strike in the summer of 1906. When some of the miners protested Zhang's decision to ignore the people in Liling County, he fired about five hundred of the protestors.[71] These fired workers represented a large percentage of the total labor force. And they now faced unemployment at a time when massive starvation had so dramatically unnerved them. The immediate termination of five hundred workers also came at about the same time that Leinung reduced the mine's labor needs by eliminating a shift of workers when he moved back to just two twelve-hour shifts from the three eight-hour shifts his German subordinate had agreed to.[72]

While conflicts erupted at the Pingxiang County Coalmines, other antigovernment forces gained strength. For example, in the countryside around Shangli where the natural disasters dramatically harmed the peasant population, a reported 90 percent of the people became members of the Gelaohui under the leadership of these men.[73] The secret society continued to gain power even after the execution of Ma Fuyi, one of its most effective leaders. After his murder in 1905, Xiao Kechang, a former military instructor, took control of the Pingxiang County lodge in his place, and under his control the secret society grew even more.[74]

As the Gelaohui expanded its power, Nationalist intellectuals from the Jiangxi and Hunanese provincial highlands tried to organize an uprising in support of Sun Yatsen. Leading intellectuals who were a part of the pro–Sun Yatsen organization, the Tongmenghui, including the Hunanese revolutionary Huang Xing, organized rebellious movements in Hunan Province during the last years of the dynasty. By 1905, the civil service exam ended and many schools and intellectual societies offered Western education. These developments led to the rise of particularly strong anti-dynastic and pro-Republican sentiments in Hunan and Jiangxi provinces.

Among the leaders of the nationalist movement in the highlands was a Japanese-trained student and member of the Tongmenghui from Pingxiang County named Cai Shaonan. Cai secretly returned to Shangli from Tokyo and made speeches calling for the overthrow of the Qing dynastic government in favor of Sun's vision of a modern republic.[75] Also leading the miners and other Pingxiang County residents was Cai's childhood friend, Wei Zongquan, the son of a wealthy merchant from Shangli and a former student in Changsha.[76] Many other intellectual reading societies and radical political organizations in Jiangxi Province came together under the Nationalist movement banner, and these organizations provided assistance to the Tongmenghui membership.[77] The two forces of the secret societies and the nationalist students merged in the spring of 1906 when several leaders of each faction agreed to form an alliance called the Hongjianghui. At a livestock fair at the village of Xiaoshui in Liuyang County, members of the Tongmenghui proposed an alliance with the Gelaohui in the interests of a unified rebellion. Under the leadership of Gong Chuntai, another Gelaohui member from the Hunanese provincial market town of Xiangtan, the students and commoners met in secret in the Mashilongwan miao behind a statue of Buddha and made a blood oath alliance. The secret meetings completed, thousands of people gathered in the village to watch operas on the newly built stages.[78] With this alliance, at least three thousand miners from the Pingxiang County Coalmines joined disgruntled miners from Shangli, factory workers from Liuyang and Liling counties, and peasants from all three counties in a general uprising under the control of a group of students allied with Sun Yatsen and the Tongmenghui.[79]

This was admittedly a loosely devised alliance with various members struggling at cross purposes. Even as some miners grew angrier about their treatment in the mines, some of the railroad workers expressed deep concern about the local Hunanese provincial government's treatment of their families after the floods. The Gelaohui's stated goal was the overthrow of the Qing dynasty and the establishment of a new, ethnic-Han empire. They argued in part that the floods and famine in the region were a sign of from Heaven that the time was ripe for a new order. Stating their complaints in more political tones, Gong Chuntai argued that the Qing Empire had failed to protect the people from foreign encroachment, had, in fact, become the collaborators of the imperialist order, and thus deserved to be overthrown.[80] Many of the nationalists, of course, dreamed of replacing the political order of the past several millennia with a republic that would put China on a new path.[81]

In the spring of 1906, local authorities conducted a series of raids on lodges and discovered in their searches that these groups planned to rebel. To stop possible violent outbursts, the Pingxiang County government bolstered the local defense organization, known as the *baojia* system, and officials temporarily halted the railroad from the Pingxiang County Coalmines to Liling County in order to restrict rebel movements.[82] At the same time, Governor-General Zhang Zhidong informed the court that a rebellion was probably imminent. Beijing quickly put together a militia from Hubei Province and sent them toward the supposed rebel area. Sheng Xuanhuai, fearing that the rebellion could do harm to his mines, sent cables to his home province of Jiangsu and the Liangjiang governor general Duan Fang ordering them to bolster the militia in the mines and the surrounding area.[83]

Even with these precautions, the government leaders' efforts were not enough to stop the Hongjianghui from carrying out their planned uprising. By the fall of 1906, the region was smoldering with rumors of imminent rebellion, even though the alliance had not yet set a date for the beginning of the uprising. Sometime in October, the Gelaohui leader Xiao Kechang reportedly told the intellectual Wei Zongquan that the rebellion needed to begin soon. He explained that the miners in the Pingxiang County coalmines planned to return to their homes for New Year's festivals. If the rebellion waited until their return, he suggested, the workers would be less likely to rebel.[84] Exactly why Xiao felt that the miners were losing their nerve was not clear, but it is obvious the leaders took his advice.

In any case, the so-called Ping-Liu-Li Uprising—referring to unrest in Pingxiang, Liuyang, and Liling Counties—began on December 7, 1906, when a reported twenty thousand poorly armed followers led by Gong

Chuntai left the village of Mashi, where they took the blood oath, and successfully attacked Shangli, placing the town under their control. Gong was a firecracker maker and a central figure in the gambling industry. He had visions of radical land and social reforms that gained the poor peasants' and workers' support. Many of his supporters in Shangli were miners who worked in the local mine shafts. They left their jobs in huge numbers that winter to return to their farms either so they could avoid the brutal conditions they experienced or as a protest of those conditions.[85] Armed with little more than the farming and mining tools that they brought with them, these rebels, wearing white bandanas and carrying white flags with the words Revolutionary Army (*Geming jun*), overwhelmed a small militia of about twenty troops stationed at the market city.[86]

Zhang Guotao was just a child of eight or nine going to school in Shangli when the city erupted in rebellion. He recalls in his memoirs that people began panicking and running in fright upon hearing of the impending rebellion. More importantly, he remembers the night that he and some of his fellow students were assaulted by the rebels who viewed Zhang as another child of the gentry class.

> It must have been about midnight when a crowd of hefty drunken men carrying sabres dragged us from bed and stood us on a counter. We awoke suddenly to see them brandishing sabres at us.
>
> "Chop off the kids' heads and wet our battle flags in their blood," some yelled. "They'd be nice to try our sabres on," bawled others. "Don't kill them," still others suggested. "Tie them up and cart them off, and let their families ransom them back with big shiny silver dollars."[87]

Once the rebel commander successfully took control of this important market city, Gong Chuntai decided to lead his followers into Liuyang County and attack the county seat. Zhang Guotao, who was somehow saved from the ravages of the mob, saw these men as they marched out of Shangli toward Liuyang County. He described them as

> a long line of shabby peasants fil(ing) in disorderly fashion down the road, carrying spears, flintlock bird guns, rakes, huge broad-bladed swords, sabers, clubs, and for shields, black iron pans and pot covers. They showed no enthusiasm: there was no flag waving or shouting as they marched toward Liuyang County.[88]

His description provides some insights as to the nature of the rebellion. The rebels were simple peasants and laborers who hoped to turn their anger toward the gentry and mine owners into some improvement in their lives. Unfortunately for them, they were not a sophisticated militia and they failed in their efforts. When Gong Chuntai reached Liuyang County, his troops were attacked by Qing forces and soundly defeated. Similarly, Liling County peasants and porcelain workers were beaten back when they attacked towns along the railroad and plundered the homes of gentry families.[89]

In the countryside, rebels moved south from Shangli along the border. As they marched, the peasants and miners used small weapons to attack the Qing troops and disrupt the transportation routes. Government forces had difficulty finding them as the mountain ridges and bamboo groves and other foliage provided the rebels with many hiding areas.[90] Even though Zhang Guotao explained that he saw no banners or revolutionary slogans carried by the rebels in Shangli, the peasants in central Pingxiang County, reportedly marching under the banner "Establish the Republic and Equalize Land Ownership," used the opportunity of the collective unrest to attack the local elites.[91] Once these rebels reached the town of Xiangdong, they turned east toward Pingxiang City where placards on the walls called on all people to assist in the overthrow of the Qing dynasty. Here they apparently planned to meet up with the workers at Anyuan in a concerted act of rebellion.

Even as peasants planned and then executed their actions throughout the countryside, inside the mining town of Anyuan itself rumors of worker unrest had spread for several months. Soldiers who attacked gambling houses and secret society halls beat and tortured men into confessing that a rebellion was being planned. Unfortunately for the Mine Bureau, Manager Zhang Zanchen was not in Anyuan—he lay dying in a hospital in Shanghai—and so a man named Lin Zhixi acted as the temporary mine manager. Lin informed Zhang Zhidong that the mines were full of rebels planning to begin a rebellion and requested extra forces to protect the Huguang governor-general's interests. To this end, the number of soldiers in Anyuan increased from one hundred to almost twice that number. Once the rebellion actually began, officials ordered more troops from Hubei and Hunan provinces to Anyuan, including an artillery force from Hunan that reportedly had 150 cannon.[92] When the troops arrived in Anyuan, they reported that they saw no rebel activity. They informed Zhang Zhidong that the miners were quiet, and therefore the militia commanders ordered their men to march to Shangli where the rebellions were most violent. Lin complained that even though the mines appeared subdued, in fact the miners were waiting until the soldiers left before they planned to rebel. Zhang Zhidong confessed to being confused by these contradictory statements and apparently allowed the soldiers to leave for the more violent areas in the north.[93]

Lin's complaints proved to be true, however, as the miners became more agitated and violent after the soldiers left the mining town. Fearing that they would be attacked by rebels, in the middle of the night Chinese Mine Bureau officials placed the foreigners in a train and sent them toward the Hunanese provincial border. Along their journey the German engineers feared that rebels had destroyed the bridge at Xiangdong. However, when they arrived at the bridge, they discovered that their fears were unwarranted and that the railroad was not harmed. They continued through Liling County and past Liling City where they saw placards calling for the overthrow of the dynasty. Finally, they arrived in the Hunanese provincial capital of Changsha where they were protected from the unrest by Chinese troops and foreign gunboats.[94]

Once the foreigners fled Anyuan, the rebels took control of the mine's administration. The miners organized under the command of a mine worker from the Bafangjin shaft, as well as three others from the Eastern Horizontal Alleyway. These men brought together about six to ten thousand miners including those five hundred who were fired in the spring.[95] They then overcame about two hundred Qing bannermen forces and Manchu officials who were encamped at a military outpost in Anyuan. They blocked up dams that flooded military outposts and destroyed the transportation routes needed by the military. Miners and Gelaohui members killed many of their opponents, leaving the rebels in a position of strength in the mine town. Those troops who were not defeated feared the rebellion and refused to try to recapture the city, leaving the mining town in the hands of the rebels. The miners used their newly acquired power in part to assist the peasants who were harmed by the spring's famine. Specifically, they forced the Hunanese provincial governor to assist the railroad workers in the industrial city of Zhuzhou as had been requested by Hunanese provincial officials more than five months earlier.[96]

The uprising included as many as twenty to thirty thousand rebels and prompted massive and swift counterattacks by Qing troops and local militia.[97] Pingxiang County's magistrate, a native of Henan Province named Zhang Zhirui, immediately summoned the Pingxiang brigade commander Hu Yinglong to fight the rebels. Hu's militia, which numbered just over 250, retook the town of Shangli. However, his forces were spread too thinly throughout the countryside and were quickly defeated by the thousands of followers of the Hongjianghui rebellion.[98] Zhang Zhidong then sent imperial armies from Changsha and Wuchang to Pingxiang County to attack the main contingents of the rebels. The forces, which reportedly included over 250,000 Hunanese troops, easily defeated the remnants of the rebellion and protected the mines from greater harm.[99]

In the meantime, various local leaders were called upon to help put down the unrest wherever they could. In the Pingxiang County Coalmines the Chinese manager Zhang Zanchen, assisted by local gentry, quickly put together a militia made up of bare sticks and *baojia* toughs.[100] Furthermore, the local *baojia* and security bureaus were deputed to rebel areas. In particular, one of the most powerful members of the *tuanlian* system was the mountain lord Li Youru. Together with Zhang Zhirui, he wrote several letters to provincial officials of the problems they had in putting down the rebellion of the peasants in the countryside.[101] Yet it is interesting that while Li was engaging in confrontations with the Pingxiang County Coalmine over control of his former mineral holdings, he was also directly involved in subduing the rebellion by miners who were struggling against the very same corporation that he opposed.

The main rebel forces were put down quickly and faced severe punishment. As one report explained: "Military law was prevailing at (Pingxiang) and decapitations were going on by the dozens day by day."[102] One gazetteer entry stated that the number decapitated in the city was between twenty and thirty, though this number may be greatly understated. The gazetteer also provides a very interesting account of the punishments of some of the rebels. It is pointed out that several of the captured leaders were members of the gentry class and that many of these were given different sentences for their crimes apparently based upon their individual actions and their status in the gentry hierarchy. While some of the leaders were executed outright, others who were judged to be of lower rank and responsibility were sentenced to death but allowed to pay a fine as redemption. Those who were members of the gentry class were given the chance to proclaim their remorse and subservience to the empire at the "Temple of the Gods" (*Shen miao*) and were subsequently stripped of their gentry status.[103]

Even as the main forces were similarly defeated, in many areas the rebels continued to engage in various acts of unrest for several weeks, if not months. One account stated that while the rebel troops were destructive toward the gentry, they were met with open arms by the "old 100 names" in village after village.[104] Li Youru, together with the Pingxiang

County magistrate Zhang Zhirui, also wrote to the Jiangxi governor of the continued unrest in Pingxiang County after the majority of the rebels were stopped. The two leaders wrote that in the mining town of Anyuan as well as Shangli and elsewhere robbers and bandits continued to pillage at night. Some estimates placed the number of rebels in Shangli alone as in the thousands.[105]

Zhang Guotao provides a brief description of these later attacks on village gentry in his memoirs. He explained that his family lived in the village of Chimushan, Pingxiang County, where their large home was attacked by the rebels. After a rebel leader assembled a mob of village supporters in front of his family's household, they slaughtered several of the family's livestock and ate in the family's quarters. The rebels did considerable damage to the home but did not harm the residents. Eventually, they left the house and moved on to other gentry homes.[106]

However, by early January of 1907 the local government pronounced that the rebellion was defeated.[107] With the leaders either executed, captured, or in hiding, government forces successfully pacified Pingxiang County. Chinese mine officials brought the German engineers back to Anyuan, and the Pingxiang County Coalmines were started up once again.[108] However, the unrest in Pingxiang County could not be entirely resolved. The struggle for better working conditions and pay inside the mines continued to supplement the peasants' demands for subsistence. More importantly, China was experiencing a much larger struggle over the future of the empire and the emperor himself. This new level of friction was added on top of the old ones and sparked yet another major uprising in the county.

The conflict between the Hanyeping Company and the local residents reached a high point in 1911 with the overthrow of the Qing dynasty. Because the revolution began in the Wuhan Cities, transportation between the Pingxiang County Coalmines and Hanyang was stopped. Furthermore, revolutionary rhetoric was in part specifically aimed at the corruption of Sheng Xuanhuai. And so, fearing that he would be killed, Sheng fled to Japan immediately after the Wuhan Uprising in October of 1911. After Sheng left Hanyang, the administration of the Hanyeping Coal and Iron Company, Incorporated, was without its leader. Since it could barely function, the firm temporarily opened an office in Shanghai during the early months after the revolution simply so it could pay its workers and tell the German engineers they were no longer needed.[109]

In Pingxiang County, the local population claimed that once the emperor left his throne, all the land rights held by the Pingxiang County Coalmines were null and void. Seizing on this opportunity, the local gentry and others quickly took control of the mines and extracted the minerals for their own purposes. Rather than maintaining the Western methods of organization and extraction, the local population apparently went back to their old patronage systems using the premodern tools and mine strategies of their past.[110] The gazetteer tells us that Bafangjin, the largest vertical mine, was abandoned and flooded, and the machinery in the mine shaft was rusted and broken.[111] One Western observer, L. C. Arlington, visited the mines while they were under the control of the local population. He explained that the mines were essentially destroyed by local misuse and neglect:

> On the outbreak of the revolution, [the German engineers] fled for their lives, leaving everything behind them, and, when I visited the place a few months afterwards, the mines were flooded to overflowing and everything was in total disorder. Tons of coal were stolen and sold in junkloads for $6 a ton; while at [Hankou], Japanese coal was being imported at $30 a ton; and while some 40,000 tons of [Pingxiang] coal were lying, not far away, at [Yuzhou], practically a drug on the market. Why some enterprising merchant did not charter a boat or steamer and take that [Yuzhou] coal to [Hankou] is an enigma.[112]

Arlington's last point that no one tried to sell the coal for huge profits to the Wuhan factories is an interesting one. When the foreigners and court officials left the mines, the local population did not try to make more money through interaction with the modern economy but instead apparently went back to their old ways of mineral extraction and sold the mineral locally as they had for centuries.[113] It would seem from this event that the local population did not adapt the ideology of the modern mines into their own culture and society. Rather, the people of Pingxiang County viewed the industrial scheme as a temporary irritant that they wished to push aside as soon as the opportunity presented itself. Even as the modern equipment lay idle and in ruins, and as China's demand for coal was as high as ever, the local population quickly returned to the lifestyles they held for centuries, producing fuel to heat their homes and fire their ovens. Just as if the previous decades had never happened, the people recreated their previous lives and ignored the chaos of the modern world that surrounded them.

Mechanization, then, acted as a hindrance to the people of the county, both rich and poor, worker and peasant, government official and commoner. Marxist scholars tell us that the Industrial Revolution led to resistance by the working classes against the mangers and overseers of industrial capitalism. The case of the Pingxiang County Coalmines certainly provides credence to that argument. The brutality of German managers and the disinterest displayed by Chinese administrators and officials sparked the coalminers at Anyuan to join into the Ping-Liu-Li Uprising. These men fought for better wages and working conditions and better treatment of their families in the countryside. Their new role as an emerging laboring force led them to join powerful secret societies that were both part of the imperial and agrarian past but, with alliances with Western-educated elites, also branched out to ideologies based in the wider world.

However, the conflicts did not end there. Local gentry also had reason to fear and oppose the mechanization as the mines separated their birthrights to land and labor in order to provide for a more expansive and efficient method of production. This new scheme was not beneficial to the lives of the gentry and therefore they sought a new dispensation. While they attempted to hinder and reduce the expansion of the mines, they also forced the Mine Bureau to fund schools required to maintain their status in the Confucian world of civil service examinations. Even more, some gentry began to see outside their immediate world as they benefited from the improved communications and transportation systems the mines provided. Young educated men had access to modern journals, and some even traveled to Japan where they studied Western thinking and organized the overthrow of the empire with Sun Yatsen.

While it is possible to view these two movements as parallel to one another, this chapter has pointed to interconnections of resistance growing out of the mechanized mines. By viewing the relationships elites and nonelites held before the coming of modern mining, it is possible to understand that the Uprising of 1906 was not only a continuation of the antagonisms these people had held for centuries, but that they were altered—indeed, modernized—in the face of a new socioeconomic and technological milieu.

In this way during the first decades of the twentieth century local conflicts and negotiations continued to alter the county while unrest throughout the empire changed China's leadership forever. And all of these fissures of conflict brought waves of change and rebirth in Pingxiang County. Even as the 1911 Revolution was meant to destroy the Confucian state, in Pingxiang County, the unrest actually provided an opportunity to return to this earlier society. Old ties between gentry and commoner were reestablished and economic connections with the outside world were severed. These conditions did not last for long, however, as the emerging Nationalist State, together with foreign assistance, returned and imposed its will on the county, furthering the mechanization scheme over the local economy for the next two decades. But even so, the glimpse of life after the 1911 Revolution was a sign that Chinese court official and foreign engineers still sat uneasily on the ground and that the centuries-old patronage system—dominated by gentry power and exploitation of commoner labor, to be sure—remained an ideal for the community. And it was this ambiguous position—teetering between various forms of the modern and the premodern—that failed to fully "self-strengthen" Pingxiang County and instead put it in a perilous and opportunistic place for the Revolutionary changes that were yet to come.[114]

Conclusion

Great Undertakings around the Globe

This book argues that some essential historical issues about industrialization are only visible at the micro-level. To modernize a coalmine, the people on the ground—nationals and foreigners, elites and nonelites, local government officials and court-appointed representatives, husbands and wives, farmers and peddlers—nearly everyone, had to be reordered and refashioned, from above and below, from within and without, into a new and interconnected community. This meant that the social status each person held when the process began almost certainly changed to accommodate the new scheme. As foreign engineers introduced new laboring schemes and new tools and devices into a community, the machinery changed the people who were charged with using that equipment. But individual communities employed industrialization schemes within the context of their own social structures, their own economic orders, and their own values. Workers were coerced or forced to change their labor strategies, their family structures, and their social and political relationships if they were to survive in the new world. Failing to do so, some peasants continued to suffer the continuing pressure of modernization on their premodern farms. Elites watched with anger as their tenants and properties were subsumed into a new economy they no longer controlled. Similarly, local officials were frustrated when ordered by their superiors to facilitate this transformation and had to either accept this commandment or leave the government and retire. In short, some people accepted the change and even prospered while others faded away in the social tumult.

At the same time, while machinery changed society, so too communities and civilizations altered the industrial process and even changed the machinery itself. Foreign experts had to rethink their own understanding of the industrialization scheme and how it could be accomplished if they hoped to modernize the community. Even as the new technology promised to increase manufacturing output, many local communities accepted these changes with some caveats and only after negotiations over working conditions, social status, and cultural mores. Foreign engineers, in turn, had to use their best diplomatic skills to entice, encourage, coerce, and force the local population to participate. They were at times able to hire strong and compliant people for one job yet forced to accept incompetent if not resistant people for another task. In those places where local practices or politics demanded, local community methods and tools often integrated with modern technology. Environmental and political conditions forced engineers to abandon even their finest equipment in order to set their "Westernization" scheme in motion.

Each community, state, and empire experienced these changes differently. While many studies of industrialization in the West suggest that mechanization led to a gradual but relentless social shift from an agrarian community based on feudal relationships to an industrialized capitalist order, I have put forward a more nuanced history in which everyone was transformed by the industrial process and the process was, in turn, altered by the actors. This history especially shows that the end result, in a decidedly brief period of a few decades, was not so much the formation of a modern economy but the breakdown of the older order.

The social and cultural changes in Pingxiang County did not solely impact the Chinese who worked in the mines but also the foreign engineers. I have shown that when Gustav Leinung arrived in Pingxiang County, he came with a set of ideas about the use of modern technology and the efficiency these tools would add to the mining process. He quickly grew frustrated with the quality of local workers, however, and subsequently scrambled around looking for Chinese men he deemed suitable to the task of implementing a Western labor and production scheme. When workers pushed back in the form of a series of labor disputes and mass actions, Leinung was forced to construct dormitories and provide other services that he hoped would satisfy the workers and quell the unrest. Most of the letters by and articles about Leinung depict a man trying to impose European efficiency on an unwilling or inept populace. But, in a letter Leinung wrote after leaving Pingxiang County, he indicated that China had also changed him. In the middle of a prospectus for another coalmining venture, Leinung wrote that he took time off from engineering and geological inquiry to do some sightseeing:

> In the afternoon of the same day, the 10th of November, it being a fine day, we still made an excursion to the temple [Taibo luo] on the [Jingting shan]. It was interesting for me

> to learn that [Li Taibo], the famous poet laureate of China, had been staying here for some while, that temple's named after him, and that he had sung an ode of the beauty of the [Jingting shan], which is a pretty one indeed.[1]

In these passages Leinung seems to suggest that his job in coalmines in China was not so much a chore as an opportunity. He did not see China simply as a backward country with workers he described at times as "lazy," but it was also an environment filled with history and wonder, a place he enjoyed and embraced. Even as some of the foreigner engineers became frustrated and returned home after a few brief years, Leinung was a welcomed and respected expert in China for more than two decades.

If Leinung's life in China made him a little less European, his work in Pingxiang County altered the miners and peasants, making them just a bit more like Leinung's European ideal and a little less like the lineage-based subsistence community they once were. I showed that prior to mechanization commoners did many different jobs to provide for their families. Miners and peasants were virtually one and the same people, or certainly members of the same community tied to lineages that provided them with safety nets for survival and social and political connections that assured them of support in their dealings with the state or their neighbors. Throughout most of the Qing era, peasant men worked the land as farmers for roughly nine months of the year and then went to the mines for the three months when farming was at its slowest to supplement income for their family. By employing divisions of labor based on gender and age, family members produced the fuel and food required for subsistence and were in turn assured of equal opportunities to consume the items of their collective production. When industrialization initially began taking over the county's coalfields, many of the miners maintained the patron-client relationships that controlled the rice fields required to feed their wives and children. Later, as men turned increasingly toward the mines for their livelihood, they severed their ties with lineage leaders, undercutting important lifelines their families had relied upon for generations. At the same time, the mechanized mines they worked extracted minerals from the same fields they previously rented from their lineage leaders in their youth, but these men were now functioning at the behest of foreign overseers and comprador Chinese managers. This meant that even if the places they worked and the brutality they endured was much as it had been before, the social ties to their ancestors and to the land their families had farmed for generations was no longer available to them for their subsistence.

When peasant men took jobs in the coalmines and became part of the industrial workforce, achieving the status of full-time laborers, the connections with the villages became increasingly tenuous. As the workers grew accustomed to living as miners, the mining town's connections with the countryside slowly dissolved. Workers stopped viewing themselves as peasants who might someday return to the villages and began to think as urban-based miners.[2] Once Leinung and his German compatriots implemented the mechanization scheme, workers entered into new relationships as skilled and unskilled labor with foreign managers and Chinese secret societies, while the social distance from their home villages and family and lineage memberships grew. This is particularly clear when viewing the Ping-Liu-Li Uprising in which the peasant forces and mining forces continued to act separately rather than as an integrated force of husbands and wives, fathers and sons joined together in resistance to their collective superiors. Later on, when the miners participated in the 1911 Revolution in Wuhan and the Nanchang and Autumn Harvest uprisings in 1927, the Communist Party took almost no notice of the peasantry in Pingxiang County. One railroad worker who worked out of Pingxiang County, Wang Zheng, argued that the peasantry did not support the Communist Party rebellions of the 1920s because they were afraid of the power of the landlords.[3] Therefore, in just two short decades, miners and peasants became separated into different classes first by the factory scheme, then the contract labor bosses and secret societies, and finally, by the organizers of the Communist Party. In each case, the overseers viewed miners as different from and separated from the villagers, even when these men did not view themselves in the same way.

The evolving relationships between miners and peasants were especially experienced in the day-to-day activities in the marketplace. For example, prior to modernization, wives and children prepared food for the husbands and adult men who came home from the mines. However, once mining expanded, peasants continued to provide daily needs to the laborers in the mines, but they did so not as mutual members of a family maintaining production portfolios for subsistence strategies but rather as merchants providing food to customers.[4] Other shops and markets sprouted up that supplemented the needs of the workers and local inhabitants. New agricultural strategies in cash crops allowed some peasants to develop new avenues of success utilizing the increased monetary wealth provided by the modernizing mine. Particularly successful peasants gained wealth that allowed them to separate from the local landlords and purchase subsistence fields for themselves. Overall however, while the Pingxiang County Coalmines provided new opportunities for some peasant families, the majority remained tenants of the landlord and gentry elites.[5]

At the same time, as the lives and positions of commoners changed under the pressure of the modern economy, the world order of the local elites turned upside down. In the years prior to industrialization, the gentry dominated the nonelite populations through patron-client agreements allowing peasants to engage in subsistence agriculture on gentry-owned property in exchange for rents required by the elites for survival and continued political and social power. The implementation of Westernization required the severance of some of those ties as the German engineer and the Chinese court took the coalmines and paddy fields from them, thus ending access to property and the labor required to make the land pay. Gentry, literally referred to as the "Lords of the Land" and the "Lords of the Mountains," were publically left powerless with the arrival of a new force that carried the visage of a foreigner and the preemptor of the court. In fact, what made this disruption even more painful was that it was designed and supported by the emperor and therefore signified the dissolution of gentry ties with the Confucian state. When the plans to change Pingxiang County's coalmining scheme began, local elites felt they had the right and duty to write of their complaints through memorials to the governing elites. However, the modernizing state ignored much of their correspondence. The industrialization of the economy along with political destruction of the Qing court meant that gentry power and status were set aside in the interests of a new idea that no longer needed them.

The same Western technology that emboldened the state also led to its ruin and harmed mutually beneficial relationships that existed between elite classes and the government. Almost immediately after the activist court forced lineage leaders to sell their lands and give up access to labor, the defeated empire ended the civil service system and gentry lost their most significant avenue to status and power. Those gentry who did succeed did so through new avenues of social advancement. For instance, in 1922, when the Communist Party member Liu Shaoqi acted as the union leader for the Anyuan workers, he sat face-to-face with local representatives of the mechanized mine and the greater Pingxiang County elite. The two leaders of this contingent were one Xie Lanfang, who was described in one Communist Party–era book as a "merchant," as well as a "local gentry" person named Zhen Shengfang.[6] Neither of the two family names Xie and Zhen held prominent places in late Qing dynasty Pingxiang County according to the 1935 gazetteer.[7] Their lineages did not include civil service degree-holders as noteworthy as men like Wen Tingshi and Li Youru. However, Xie became an important person in local politics and the economy through a new avenue not available to those other gentry leaders. He used his expertise as a merchant to attain the position of director of the Anyuan Chamber of Commerce. And with this achievement, he held significant influence on the politics of the time, even inadvertently assisting the Communist leaders Li Lisan and Liu Shaoqi in their endeavors to organize the workers under their union banner.[8] The rise of men such as Xie and Zhen, suggests both a transformation of the leaders of Pingxiang County during the Republican period as well as a marginalization of the old leadership in favor of a new generation of merchant-oriented leaders.[9]

In the same way that the mechanization of the coalmines and the collapse of the imperial government changed the status of local gentry, so too strategies for success in the government were altered and expanded. Prior to mechanization, magistrates oversaw Pingxiang County as representatives of the imperial court. Men like Gu Jiaxiang were Confucian scholars who viewed their jobs as integral to both the success of the court and the correctness of the state-sanctioned religion. Among the most important duties of the magistrates for the previous centuries were negotiations between elites and nonelites over such issues as rental agreements, labor output, and sales in the marketplace. Gu and his predecessors worked with local gentry to construct bridges and orphanages and promoted the schools in the hopes of further success in the civil service system. Yet when Sheng Xuanhuai commanded Gu to assist in mechanizing Pingxiang County, Gu had to use whatever powers he held to take land and wealth away from those same gentry leaders. Court-appointed officials and merchants who arrived to fulfill a myriad of duties required for industrialization further overwhelmed Gu. Chinese leaders from throughout the empire who in the past would have been content to act as local gentry in their own counties sought new avenues of status and power developing skills as linguists, engineers, managers, and merchants. From exemplars of Confucian virtue, such officials took on new jobs as subordinates of the modern order who no longer sought the active support of their fellow gentry leaders. When the empire collapsed in 1911, the industrial plants in Pingxiang County and elsewhere continued to function, though they were increasingly overseen by government officials even less well suited than their predecessors. The local yamen designed to be ruled by scholarly gentlemen came under the control of warring factions and warlords whose voracious desire for local wealth corrupted the political scheme.[10]

It was the young students, the men who allied with peasants and miners in the Ping-Liu-Li Uprising, who ultimately transformed the county and the empire. Some of these nationalists were linked to Sun Yatsen while others gained status as union organizers and revolutionaries within his Guomindang Party. These men, with the continued help of the peasants and commoners of China, moved China on a new path. Their final goal was not, however, the furtherance of industrial capitalism like their Western counterparts. Instead, Mao Zedong and other young leaders of the coalmine labor movement in 1920s Pingxiang County called on their followers to turn away from this foreign definition of modernization. And for several decades China abandoned the industrial skills, machinery, and trappings of modernity that increasingly dominated Pingxiang County in the late nineteenth and early twentieth centuries.

Indeed, I have shown that many of the forces that brought about the Communist Revolution in China were vividly illustrated at the micro level in Pingxiang County: Westernization, the dissolution of patronages, and political disintegration. Pingxiang County's experience was neither the most significant nor especially common. What makes this case important is that mining communities underwent these changes more rapidly than almost anywhere else. Unlike the case I have presented of mechanized mining in Pingxiang County in the late nineteenth century, Western influence first arrived as early as the sixteenth century, initially coming in the form of military might and dogged marketing. This early invasion primarily centered on Chinese ports, markets, and provincial capitals that were already developed by centuries of regional trade and political influence. Westerners purchased and sold items in the city markets, and Chinese merchants and workers in turn plied their trades for the intruders much as they had supplied internal markets for generations. Over the next several decades, Chinese and Westerners slowly developed, negotiated, and expanded their relationships in these cities and ports. In these cities, Chinese were gradually transformed as influence slowly grew.[11]

On the other hand, for the people of Pingxiang County, the transformation was sudden and dramatic. On the eve of modernization elites and nonelites alike lived in a subsistence economy located on the frontier of China's political and economic realm. In the fall of 1896, some peasants who harvested their crops as their ancestors had for hundreds of years were subsequently hired by court officials and the German engineer

Gustav Leinung, even as the winter wheat was maturing in the fields. During these initial stages, local gentry attempted to negotiate pacts with the state that allowed for continuation of the social structures that legitimized and solidified their power while putting out the possibility of securing greater profits in the future. Mine laborers and transportation families attempted to integrate their skills and previous efforts into the new structure being put in place. However, the German engineers quickly determined that the local workers and their tools and methods of extraction were insufficient. These outsiders imported machinery and laborers from outside Pingxiang County and engineers from Germany. At the same time, Chinese officials were ordered by the court to construct a modern mine and railroad in the interests of the empire knowing full well that they might harm the lives of the local population. They helped the German expert create a modern mining city that employed scientific and technological advancement far beyond that of any Chinese-sponsored firm. Yet, even to the degree that the city of Anyuan was growing into a mini version of Shanghai or Beijing, London or Bonn, in reality it was little more than an overgrown factory that could not hope to provide the same legal or social services provided by the lineages and county magistrates of the past. Nor could it promise the equivalent level of economic gains enjoyed by inhabitants of other larger and more developed Chinese treaty ports let alone fully developed European cities.

Of course, it may seem obvious that technology and know-how alone cannot bring about the so-called take-off needed for any country's successful transformation. The industrialization of a country or empire required more than the desire and skills of the state and its most powerful and connected people, and more than the employment and imposition of Western technology and all its promises. Mechanization also necessitated dramatic transformations of local politics and society on the ground, often in the least-developed communities of the state. To fully understand the requirements of industrialization, we must examine more closely these political changes and see the unintended consequences that transpire among the local elites and nonelites alike. Modernization is not simply a process at the macro level but often requires a myriad of specific changes in communities in the sites where mineral resources are found, usually far away from the natural ports and trading posts and much higher in elevation than the luxurious crop-filled lowlands. In countries and empires throughout the world, reformers and imperialists were primarily concerned with the capital and machinery of their industrial scheme, but the local and individual component was an equally significant aspect in the history of modernization. In fact, long after the machinery rusted away and the capital investments were lost or recouped, the experiences and ideas, the bonds and conflicts, lived on in the community in the guise of such organizations as the Communist Party in China and other labor and political movements around the globe. In essence, the scheme that was literally supposed to provide China with an empire-wide "self-strengthening" actually atomized the population and undermined its unity.

As I have shown, rapid and dramatic mechanization led to both explosive economic growth and crippling social devolution. And this is as true today as it was then. Technology transfer is not one experience and it cannot be negotiated by simply integrating a new set of social needs within a predetermined ancien régime. Elites and nonelites hold different relations in each empire, country, state, and village in the world. Mechanization integrates, transforms, and destroys those social ties in different ways from one place or one set of economic needs and conditions to the next. How this was accomplished from place to place altered the future landscape created by the new order. And in this world, men and women found new lives, occupations, and ties to the larger community and world. Some people gained much from the changes being attempted in the county, taking jobs and developing skills required by the new order. Others begrudgingly accepted their new lives, attempting to do what was necessary to maintain their former ways of life within the confines of the new order. Still others organized committees, unions, and religious and political orders that they hoped would end the new order. Rather than accepting or embracing the dramatic changes they saw occurring in their communities, these people fought like hell to make them stop.

Notes

Abbreviations

ACDC: American China Development Company. *Contracts, Chinese Government and American China Development Company*: Dated April 14th, 1898, and July 13th, 1900. China: American China Development Company, 1900–1904.

AYBGSL: Jiangxi sheng Pingxiang shi zong gong hui and Anyuan lukuang gongren yundong ji hui guang, eds. *Anyuan lukuang gongren dabao shengli liushi zhou nian ji hui hua ce* (Anyuan Railroad and Mining Labor Strike Victory Sixtieth Anniversary Commemorative Photo Collection). Jiangxi: N.p., 1982.

AYLK: Changsha shi geming jinian di bagong tai and Anyuan lukuang gongren yundong jinian guang, eds. *Anyuan lukuang gongren yundong shiliao* (Historical materials of the Anyuan Railroad and Mining Labor movement). Changsha: Hunan gongren chubanshe, 1980.

Contracts: Pingxiang mei ju. "Pingxiang meikuang chanye qi" (Contracts of the Pingxiang County industries). Unpublished.

HAD: Zhonggong Pingxiang meikuang weiyuan hui xuan zhuan bubian, eds. *Hongse de Anyuan* (Red Anyuan). Nanchang: Jiangxi renmin chubanshe, 1959.

KWD: Zhong yang yanjiuyuan and Jindai shi yanjiusuo. *Kuang wu dang* (Archives of Mining Matters) Vol. 4: Anhui, Jiangxi, Hubei, and Hunan provinces. Taibei: Zhong yang yanjiuyuan jindai shi yanjiusuo, 1960.

JDMKS: Zhongguo mindai meikuang shi, eds. *Zhongguo mindai meikuang shi* (Modern coalmining affairs in China). Beijing: Meikuang gongye chubanshe, 1990.

JDSZL: Zhongguo kexueyuan jindaishi yanjiusuo shiliao bianyizu, eds. *Jindai shi ziliao* (Modern history materials). Beijing: Beijing kexue chubanshe. Irregular Periodical.

LSQYA: Zhongguo shehui kexueyuan jindaishi yanjiusuo and Anyuan gongren yundong jinianyuan, eds. *Liu Shaoqi yu Anyuan gongren yundong* (Liu Shaoqi at the Anyuan Labor movement). Beijing: Zhongguo jinian yuanxue, 1981.

PKJS: Pingxiang kuang wu ju, eds. *Jinian ce: Ping kuang jianshe jiushi zhouhua* (A commemorative volume: The ninetieth anniversary of the establishment of the Pingxiang Mine). Pingxiang: Ping kuang gongren baoshe yinshuachang, 1988.

PKWJ: Pingxiang kuangwu ju zhi weiyuan hui, eds. *Pingxiang kuangwu ju zhi* (Documents on the Pingxiang mining affairs bureau). Jiangxi: Pingxiang kuangwu ju zhi bian weiyuan hui, 1988.

PLLQY: Pingxiang shi zheng xie, Liuyang xian zheng xie, and Liling shi zheng xie, eds. *Ping-Liu-Li qiyi ziliao huibian* (Compilation study of the Ping-Liu-Li Uprising). Changsha: Hunan renmin chubanshe, 1986.

PXRW: Pingxiang wenshi ziliao (7th division) and Pingxiang shi zhi tong xun (14th edition), eds. *Pingxiang renwu ji lue* (Brief Records of the Activities of People from Pingxiang). Pingxiang: Pingxiang shi zheng xie wenshi ziliao yanjiu weiyuan hui and Pingxiang shi zhi bian luo weiyuan hui, 1987.

PXSLS: Pingxiang shi liangshi ju. *Pingxiang shi liangshi zhi* (Documents on Pingxiang City grain). Nanchang: Jiangxi renmin chubanshe, 1992.

XHGM: Zhongguo shixuehui, ed. *Xinhai geming* (1911 Revolution materials). Shanghai: Renmin chubanshe, 1957.

YWYD: Zhongguo shi xuehui, Zhongguo ke xueyuan, and Zhong yang danganguan, eds. *Yangwu yundong* (Western Affairs movement), 8 Vols. Shanghai: Shanghai renmin chubanshe and Shanghai shu dian chubanshe, 2000.

Introduction

1. Anthony F. C. Wallace, *St. Clair: A Nineteenth-Century Coal Town's Experience with a Disaster-Prone Industry* (New York: Knopf, 1987); and Saleem H. Ali, *Mining, the Environment, and Indigenous Development Conflicts* (Tucson: University of Arizona Press, 2003).

2. Albert Feuerwerker, "China's Nineteenth-Century Industrialization: The Case of the Hanyeping Coal and Iron Company, Inc.," in *The Economic Development of China and Japan: Studies in Economic History and Political Economy*, ed. C. D. Cowan (New York: Frederick A. Praeger, 1964), and Albert Feuerwerker, *China's Early Industrialization: Sheng Hsuan-huai (1844–1916) and Mandarin Enterprise* (New York: Atheneum, 1970); Wellington K. K. Chan, *Merchant*

Mandarins and Modern Enterprise in Late Ch'ing China (Cambridge: Harvard University Press, 1975), and Wellington K. K. Chan, *Politics and Industrialization in Late Imperial China* (Singapore: Institute of Southeast Asian Studies, 1975); Elisabeth Köll, *From Cotton Mill to Business Empire: The Emergence of Regional Enterprises in Modern China* (Cambridge: Harvard University Press, 2003). For a slightly different study of Western engineers in China, see Wu Shellen Xiao, "Underground Empires: German Imperialism and the Introduction of Geology in China, 1860–1919," PhD dissertation (Princeton University, 2010).

3. Michael T. Taussig, *Devil and Commodity Fetishism in South America* (Chapel Hill: University of North Carolina Press, 1980); Joe William Trotter, *Coal Class and Color: Blacks in Southern West Virginia, 1915–32* (Urbana: University of Illinois Press, 1990).

4. Lynda Shaffer, "Anyuan: The Cradle of the Chinese Workers' Revolutionary Movement, 1921–22," in *Columbia Essays in International Affairs*, ed. Andrew W. Cordier (New York: Columbia University Press, 1969), 166–201, and Lynda Shaffer, *Mao and the Workers: The Hunan Labor Movement, 1920–1923* (Armonk: M. E. Sharpe, 1982); Elizabeth J. Perry, *Anyuan: Mining China's Revolutionary Tradition* (Berkeley: University of California Press, 2012).

5. Zhonggong Pingxiang meikuang weiyuan hui xuan zhuan bubian, eds., *Hongse de Anyuan* (Nanchang: Jiangxi renmin chubanshe, 1959).

6. Tim Wright, ed., *The Chinese Economy in the Early Twentieth Century: Recent Chinese Studies* (New York: St. Martin Press, 1992); Xu Dixin and Wu Chengming, eds., *Chinese Capitalism, 1522–1840* (Houndmills: Macmillan, 2000); Tim Wright, "'The Spiritual Heritage of Chinese Capitalism': Recent Trends in the Historiography of Chinese Enterprise Management," *The Australian Journal of Chinese Affairs* 19, 20 (1988): 185–214.

7. Chen Xulu and Gu Tinglong, eds., *Hanyeping gongsi*, Vols. I and II (Shanghai: Shanghai renmin chubanshe, 1984 and 1986).

Chapter 1. Scratching the Dirt, Digging the Rocks: The Economy and Technology of Late Imperial Era Pingxiang County

1. For a "macroregional" map of the area, see G. William Skinner, "Presidential Address: The Structure of Chinese History," *Journal of Asian Studies* 44, 2 (1985): 273. In this article Skinner separated Jiangxi from Hunan and Hubei provinces.

2. Ren Mei'e, Yang Renzhang, and Bao Haosheng, *An Outline of China's Physical Geography*, trans. Zhang Tingquan and Hu Genkang (Beijing: Foreign Languages Press, 1985), 29.

3. Liu Hongpi, ed., *Zhaoping zhilue* (Taiwan: Chengwen chubanshe, 1975), 40–41, 48–51; Huang Liu-hung, *A Complete Book Concerning Happiness and Benevolence: A Manual for Local Magistrates in Seventeenth-Century China*, trans. and ed. Djan Chu (Tucson: University of Arizona Press, 1984), 120–21;

Piper Rae Gaubatz, *Beyond the Great Wall: Urban Form and Transformation on the Chinese Frontiers* (Stanford: Stanford University Press, 1969), 323.

4. Walworth Tyng, "The Miners' Church at Peaceful Spring: Among Collieries and Coke Ovens at Anyuen—A Vivid Picture of Our Work in a Little Known Part of the District of Hankow," *The Spirit of the Missions* 90 (1925): 477.

5. Liu, *Zhaoping zhilue*, 78, 2272–73.

6. Liu, *Zhaoping zhilue*, 212.

7. Tyng, "Miner's Church at Peaceful Spring," 478.

8. Liu, *Zhaoping zhilue*, 79–80.

9. Liu, *Zhaoping zhilue*, 78.

10. Joseph W. Esherick, *Reform and Revolution in China: The 1911 Revolution in Hunan and Hubei* (Berkeley: University of California Press, 1976), 62; Chang Kuo-t'ao, *The Rise of the Chinese Communist Party, 1921–1927: The Autobiography of Chang Kuo-t'ao* (Lawrence: University of Kansas Press, 1971), 5–6.

11. Liu, *Zhaoping zhilue*, 51–52, 78; Fu Xiongxiang, *Liling xiangtu zhi* (Taiwan: Chengwen chubanshe, 1926), 24–26; Huang Zuxun, *Luyang xiangtuzhi* (Taibei: Qingshi yingshu chubanshe, 1967), 251.

12. A. V. Chayanov, *The Theory of Peasant Economy* (Madison: University of Wisconsin Press, 1986), 94.

13. Philip C. C. Huang, *The Peasant Economy and Social Change in North China* (Stanford: Stanford University Press, 1985), 7.

14. John Lossing Buck, *Land Utilization in China: An Atlas* (Shanghai: University of Nanking Press, 1937), 37, 86–91.

15. William Barclay Parsons, "From the Yang-Tse Kiang to the China Sea," *The Geographical Journal* 19, 6 (1902): 732; "The Pingsiang Colliery: A Story of Early Mining Difficulties in China," *The Far Eastern Review* 12, 10 (1916): 375. For more on this, see Francesca Bray, *Science and Civilization in China, Volume 6: Biology and Biological Technology, Part II: Agriculture* (Cambridge: Cambridge University Press, 1984), and Francesca Bray, *Rice Economies: Technology and Development in Asian Societies* (Oxford: Basil Blackwell, 1986), 36–37, 147–69.

16. Chen Xulu and Gu Tinglong, eds., *Hanyeping gongsi*, Vol. 1 (Shanghai: Shanghai renmin chubanshe, 1984), 110.

17. Liu, *Zhaoping zhilue*, 2368–69.

18. Buck, *Land Utilization*, 86–91.

19. Liu, *Zhaoping zhilue*, 2368–72.

20. Liu, *Zhaoping zhilue*, 2368–72.

21. Buck, *Land Utilization*, 50–66, 102.

22. Ren et al., *China's Physical Geography*, 232–33.

23. For more descriptions of the importance of sweet potatoes on Chinese production schemes of the late nineteenth century, see Huang, *Peasant Economy and Social Change*, 109, 116–17, and Sucheta Mazumdar, *Sugar and Society in China: Peasants, Technology and the World Market* (Cambridge: Harvard University Press, 1998).

24. Dwight Perkins, *Agricultural Development in China, 1368–1968* (Chicago: Aldine Press, 1969), 48–51; Ho Ping-ti, *Studies on the Population of China, 1368–1953* (Cambridge: Harvard University Press, 1959), 184–90. The 1873 gazetteer of Pingxiang County does not list sweet potatoes among the crops found in the county. However, the later gazetteer of 1935 describes the planting of sweet potatoes as an important crop. See Liu, *Zhaoping zhilue*, 871. And Buck's survey finds that while it was not important in the northern, more developed, part of the county, in the south between 10 and 19 percent of the land was under cultivation of the New World crop. See Buck, *Land Utilization*, 82–83.

25. For more on women weaving and performing other domestic tasks, see Francesca Bray, *Technology and Gender: Fabrics of Power in Late Imperial China* (Berkeley: University of California Press, 1997).

26. Thomas T. Read, "The Mineral Production and Resources of China," *Transactions of the American Institute of Mining Engineers* 43 (1912): 296.

27. Philip C. C. Huang, *The Peasant Family and Rural Development in the Yangzi Delta, 1350–1988* (Stanford: Stanford University Press, 1990), 49–57. For more on the role of family strategies in rice agriculture, see Bray, *Agriculture*, and Clifford Geertz, *Agricultural Involution: The Processes of Ecological Change in Indonesia* (Berkeley: University of California Press, 1963).

28. Liu, *Zhaoping zhilue*, 1721–22.

29. Buck, *Land Utilization*, 116; Liu, *Zhaoping zhilue*, 385.

30. Liu, *Zhaoping zhilue*, 900. In fact, Pingxiang County is located between two porcelain centers, the Liuyang and Liling counties to the immediate west and China's greatest porcelain center in Jingdezhen to the east.

31. Liu, *Zhaoping zhilue*, 892–99; P. Kao and H. C. Hsu, "Geology of Western Kiangsi," *Geological Memoirs* 16 (1940): 69; Esherick, *Reform and Revolution*, 60–65.

32. V. K. Ting, "China's Mineral Resources," *The Far Eastern Review* 15, 2 (1919): 80; Kao and Hsu, "Geology of Western Kiangsi," 66; Wilfred Smith, *A Geographical Study of Coal and Iron in China* (Liverpool: University Press of Liverpool, 1926), 11, 18.

33. Liu, *Zhaoping zhilue*, 892–93.

34. Hua Wen and Luo Xiao, "Pingxiang meitan fazhan," in *Pingxiang mietan fazhan shilue*, ed. Jiangxi shegn zheng xi wenshi ziliao weiyuan hui and Pingxiang shi zheng wei wenshi ziliao yanjiu weiyuan hui he bian (Hong Kong: N.p., 1987), 1.

35. In his research on Song dynastic industrial capacity in northern China, Hartwell argues that by the eleventh century the technological sophistication of ironworks and the manufacture and development of other metals and supporting minerals was equal to Europe's at the beginning of the Industrial Revolution. See Robert Hartwell, "Cycle of Economic Change in Imperial China: Coal and Iron in Northeast China, 750–1350," *Journal of the Economic and Social History of the Orient* 10 (1967): 154.

36. Hua and Luo, "Pingxiang meitan fazhan," 1.

37. Changsha shi geming jinian di bagong tai and Anyuan lukuang gongren yundong jinian guang, eds., *Anyuan lukuang gongren yundong shiliao* (Changsha: Hunan gongren chubanshe, 1980), 462.

38. Luo Xiao, *Pingxiang shi difang meitan gongye zhi* (Nanchang shi: Jiangxi renmin chubanshe, 1992), 47–48; Alfred C. Reed, "Coal Mining in China," *The Scientific Monthly* 5 (1917): 47–49.

39. Smith, *Geographical Study of Coal and Iron*, 38.

40. Reed, "Coal Mining in China," 47.

41. Rudolf P. Hommel, *China at Work: An Illustrated Record of the Primitive Industries of China's Masses, Whose Life Is Toil, and Thus an Account of Chinese Civilization* (New York: John Day, Company, 1937), 3.

42. Ferdinand Paul Wilhelm Freiherr von Richthofen, *Baron Richthofen's Letters, 1870–1872* (Shanghai: North China Herald, n.d.), 6.

43. Reed, "Coal Mining in China," 44.

44. Hommel, *China at Work*, 2–11; Bray, *Rice Economies*, 48–50; Bray, *Agriculture*, 318; Peter Perdue, *Exhausting the Earth: State and Peasant in Hunan* (Cambridge: Harvard University Press, 1987), 127–30.

45. Luo, *Pingxiang shi difang meitan*, frontispiece; Peter J. Golas, *Science and Civilization in China: Volume 5: Chemistry and Chemical Technology, Part XII: Mining* (Cambridge: Cambridge University Press, 1999), 260–78.

46. Boris Torgasheff, *Mining Labor in China* (Shanghai: Bureau of Industrial and Commercial Information, Ministry of Industry, Commerce and Labor, and National Government of the Republic of China, 1930), 14, 163.

47. Golas, *Mining*, 320, 340–41; Leonard G. Ting, "The Coal Industry of China," Pt. I, *Nankai Social and Economic Quarterly* 10, 1 (1937): 68; Hommel, *China at Work*, 2–3.

48. Hommel, *China at Work*, 2–9.

49. Golas, *Mining*, 288–300; Ting, "Coal Industry of China," 34.

50. Hommel, *China at Work*, 4; Tim Wright, *Coal Mining in China's Economy and Society, 1895–1937* (Cambridge: Cambridge University Press, 1984), 36.

51. Golas, *Mining*, 344.

52. Reed, "Coal Mining in China," 47–48.

53. Kenneth Pomeranz, *The Great Divergence: China, Europe and the Making of the Modern World Economy* (Princeton: Princeton University Press, 2000), 65–66. His thesis is particularly important here as he goes on to suggest that since few mines experienced problems with water damage and seepage that the pump—one of the most important inventions that sparked the Industrial Revolution—was not developed by the Chinese. It is curious that, since Pomeranz argues that ventilation was important in Chinese mines, he did not discuss these similar devices. Either he could argue that ventilation ducts were not developed—which is not true—or he could show that ventilation might have sparked technology needed for water pumps or other important inventions of the Industrial Revolution, a counterhypothesis similar to his theories of water pumps that could be equally true.

54. Golas, *Mining*, 336; Wright, *Coal Mining in China's Economy and Society*, 5.

55. Reed, "Coal Mining in China," 46; N. Berkowitz, *An Introduction to Coal Technology* (New York: Academic Press, 1979), 223–24; C. C. Hsiao, "Criteria for the Evaluation of Blast-Furnace Coke," *Geological Bulletin* 30 (1937): 77–79.

56. Smith, *Geographical Study of Coal and Iron*, 38.

57. Chen and Gu, *Hanyeping gongsi*, Vol. 1, 259–60; Richthofen, *Baron Richthofen's Letters, 1870–1872*, 5; Liu, *Zhaoping zhilue*, 1549–51; Gu Lang, *Zhongguo shi da kuang chang diaocha ji* (Shanghai: Commercial Press, 1916), chap. 3, 16–17; T. F. Hou, *General Statement on the Mining Industry*, Vol. 5 (Beijing: Ministry of Agriculture and Mines, 1929), 479.

58. Ting, "Coal Industry of China," 36; William Barclay Parsons, *Railways in China* (Philadelphia: Engineers of Philadelphia, 1916), 37.

59. Richthofen, *Letters*, 6.

60. Thomas G. Rawski, *Economic Growth in Prewar China* (Berkeley: University of California Press, 1989), 184. See also Ting, "Coal Industry of China," 39.

61. Wright, *Coal Mining in China's Economy and Society*, 31–32.

62. For more on wages, see Madeline Zelin, *Merchants of Zigong: Industrial Entrepreneurship in Early Modern China* (New York: Columbia University Press, 2005), 91–92, 124–25; Sun E-tu Zen, "Mining Labor in the Ch'ing Period," in *Approaches to Modern Chinese History*, ed. Albert Feuerwerker (Berkeley: University of California Press, 1967), 62; Sidney Gamble, "Daily Wages of Unskilled Chinese Laborers, 1807–1902," *Far East Quarterly* 3, 1 (1943): 43; Mao Zedong, *Report from Xunwu*, trans. with an intro. by Roger R. Thompson (Stanford: Stanford University Press, 1990), 222; Huang, *Peasant Economy and Social Change*, 59. Conversion rates are notoriously imperfect. For some suggestions on these calculations, see Frederick Wakeman Jr., *The Great Enterprise: The Manchu Reconstruction of Imperial Order in Seventeenth-Century China* (Berkeley: University of California Press, 1985), xiii; and Jerome Ch'en, *The Highlanders of Central China: A History, 18950–1937* (Armonk: M. E. Sharpe, 1992), xix–xxii.

63. Hua and Luo, "Pingxiang meitan fazhan," 2

64. G. William Skinner, "Marketing and Social Structure in Rural China," Part II, *Journal of Asian Studies* 24, 1 (1965): 195–205.

65. Cited in Golas, *Mining*, 351.

66. G. James Morrison, "Journeys in the Interior of China," *Proceedings of the Royal Geographical Society and Monthly Record of Geography* 2, 3 (1880): 158.

67. William Rowe, *Hankow: Commerce and Society in a Chinese City, 1796–1889* (Stanford: Stanford University Press, 1984), 295; Richthofen, *Letters*, 4.

68. Ting, "Coal Industry of China," 35. For one family that might have had luck selling to the people of Hunan, see Esherick, *Reform and Revolution*, 58.

69. Rose Kerr and Nigel Wood, *Science and Civilization in China: Volume 5: Chemistry and Chemical Engineering, Part XII: Ceramic Technology*, ed. Rose Kerr with contributions from Ts'ai Me-fen and Zhang Fukang (Cambridge: Cambridge University Press, 2004), 316–18, 533–34.

70. Chen and Gu, *Hanyeping gongsi*, Vol. 1, 110, 136.

71. George G. Chisholm, "The Resources and Means of Communication in China," *The Geographic Journal* 12, 5 (1898): 50; Ting, "Coal Industry of China," 37–38.

72. Parsons, "From the Yangtse Kiang," 714. See also, Richthofen, *Letters*, 1; Albert S. Bickmore, "Sketch of a Journey from Canton to Hankow," *Journal of the Royal Geographical Society of London* 38 (1868): 54.

73. Richthofen, *Letters*, 1.

74. Morrison, "Journeys in the Interior of China," 154–55.

75. Parsons, "From the Yangtse Kiang," 715.

76. Morrison, "Journeys in the Interior of China," 160. However, not all of the travelers were so sanguine, as subsequent treks by Western observers provided less optimistic calculations even as they confirmed the descriptions of active coal markets along the Xiang River basin. On this point, see W. Clayton Grosevernor, "The Province of Hunan: Some Characteristics and Peculiarities," *Scottish Geographical Magazine* 44 (1928): 144–50. See also Percy M. Roxby, "Wu-Han: The Heart of China," *Scottish Geographical Magazine* 32 (1916): 266–79; and Ting, "Coal Industry of China," 47–50.

77. Richthofen, *Letters*, 6.

78. Richthofen, *Letters*, 10; Chisholm, "The Resources and Means of Communication," 505.

79. Grosevernor, "The Province of Hunan," 145.

80. Richthofen, *Letters*, 6–7; Ting, "Coal Industry of China," 65.

81. Richthofen, *Letters*, 4, 6, 10; Chisholm, "The Resources and Means of Communication," 501–02.

82. Parsons, "From the Yangtse Kiang," 724.

83. Parsons, "From the Yangtse Kiang," 734.

84. For a similar argument, see Ting, "Coal Industry of China," 37.

85. Pomeranz, *Great Divergence*, 63–64.

86. Golas, *Mining*, 188. Elspeth Thompson puts the amount of proven coal deposits at 1,001.9 billion tons today. See Elspeth Thompson, *The Chinese Coal Industry: An Economic History* (London: Routledge Curzon, 2003), 1.

87. Golas, *Mining*, 188. For provincial output figures for most of China, see Read, "Mineral Production and Resources of China," 297. For provincial output figures for the late nineteenth and early twentieth centuries, see Wright, *Coal Mining in China's Economy and Society*, 10.

88. Ting, "Coal Industry of China," 51–52.

89. Philip C. C. Huang, "Development or Involution in Eighteenth-Century Britain and China? A Review of Kenneth Pomeranz's *The Great Divergence: China, Europe and the Making of the Modern World Economy*," *Journal of Asian Studies* 61, 2 (2002): 533; Wright, *Coal Mining in China's Economy and Society*, 10–12. Based on estimates from the League of Nations, Ting lists the 1925 per capita consumption of coal as .059 for China, 3.9 for England, and 4.4 for the United States; see Ting, "Coal Industry of China," 207.

90. Xue Yong, "'Fertilizer Revolution'? A Critical Response to Pomeranz's Theory of Geographic Luck," *Modern China* 33, 2 (2007): 218–22.

91. Shannon R. Brown and Tim Wright, "Technology, Economics, and Politics in the Modernization of China's Coal-Mining Industry," *Explorations in Economic History* 18, 1 (1981): 63.

92. Ting, "Coal Industry of China," 205.

93. This discussion has led to an important debate between Pomeranz and some of his critics. For some of this back-and-forth, all of which included discussions of Pomeranz's arguments concerning coal and coalmining, see Huang, "Development or Involution"; and Philip C. C. Huang, "Further Thoughts on Eighteen-Century Britain and China: Rejoinder to Pomeranz's Response to My Critique," *Journal of Asian Studies* 62, 1 (2003): 157–67; as well as Kenneth Pomeranz, "Beyond the East-West Binary: Resituating Development Paths in the Eighteenth-Century World," *Journal of Asian Studies* 61, 2 (2002): 530–90, and Kenneth Pomeranz, "Facts Are Stubborn Things: A Response to Philip Huang," *Journal of Asian Studies* 62, 1 (2003): 167–81.

Chapter 2. Relatives, Clansmen, and Neighbors: Local Politics on the Eve of Mechanization

1. Robert Brenner, "The Agrarian Roots of European Capitalism," *Past and Present* 97 (1982): 30–75, and Robert Brenner, "The Social Basis of Economic Development," in *Analytical Marxism*, ed. John Roemer (Cambridge: Cambridge University Press, 1986); Kathy Le Mons Walker, *Chinese Modernity and the Peasant Path: Semicolonialism in the Northern Yangzi Delta* (Stanford: Stanford University Press, 1999); Sucheta Mazumdar, *Sugar and Society in China: Peasants, Technology, and the World Market* (Cambridge: Harvard University Press, 1998); Christopher Mills Isett, *State, Peasant, and Merchant in Qing Manchuria, 1644–1862* (Stanford: Stanford University Press, 2007).

2. Brenner, "Agrarian Roots," and Brenner, "Social Basis of Economic Development."

3. Walker finds that Jiangnan China's social and economic systems are similar to Brenner's description of France. Specifically, she argues that while the landlords surrounding the city of Nantong continued to hold certain political and legal constraints on commoners, they were incapable of capturing labor power or of demanding rents or the right to replace peasant families who worked their properties. Instead, the peasant population controlled the means of production and thus output of their fields. See Walker, *Chinese Modernity and the Peasant Path.*

4. Robert Brenner, "The Origins of Capitalist Development: A Critique of Neo-Smithian Marxism," *New Left* 104 (1977): 25–92.

5. Peter Perdue, *Exhausting the Earth: State and Peasant in Hunan* (Cambridge: Harvard University Press, 1987), 72–77.

6. Stephen Carl Averill, "Revolution in the Highlands: The Rise of the Communist Movement in Jiangxi Province," PhD dissertation (Cornell University, 1982), 8.

7. Quoted in Ho Ping-ti, *Studies on the Population of China, 1368–1953* (Cambridge: Harvard University Press, 1959), 146.

8. Averill, "Revolution in the Highlands," 158; Liu Hongpi, ed., *Zhaoping zhilue* (Taiwan: Chengwen chubanshe, 1975), 503–784.

9. Liu, *Zhaoping zhilue*, 791. In an article on official population numbers, Skinner proves convincingly that county population data tended to depict population increase because unscrupulous local officials simply invented the numbers to legitimate their performances. He showed that the figures in the years after a new emperor took the throne tended to be more realistic as the local officials were sternly warned that they were to conduct a population survey and provide proper numbers. Subsequent entries were less accurate. See G. William Skinner, "Sichuan's Population in the Nineteenth Century: Lessons from Disaggregated Data," *Late Imperial China* 8, 1 (1987): 1–79. For another study of this problem, see Ho, *Studies on the Population*, in particular chapter 1.

10. See, for example, Liu, *Zhaoping zhilue*, 2281–82. These names for migrants are terms of derision, yet they have been maintained by scholars as the best methods of identifying the peoples who arrived in the Jiangxi provincial highlands during the Qing dynasty, and they will be referred to by these names in this study. Averill, "Revolution in the Highlands," 9, 11, 13; Ho, *Studies on the Population*, 145; Anne Rankin Osborne, "Barren Mountains, Raging Rivers: The Ecological and Social Effects of Changing Land Use in the Yangzi Periphery in Late Imperial China," PhD dissertation (Columbia University, 1989), 8; Sow-Theng Leong, *Migration and Ethnicity in Chinese History: Hakkas, Pengmin, and Their Neighbors*, ed. Tim Wright (Stanford: Stanford University Press, 1997).

11. Averill, "Revolution in the Highlands," 22; Fu Xiongxiang, *Liling xiangtu zhi* (Taiwan: Chengwen chubanshe, 1926), 66–67. For more on the prerevolutionary timber industry in Pingxiang County, see Huang Shigao, ed., *Pingxiang shi zhi* (Pingxiang: Pingxiang shi zhi bianzuan weiyuan huibian gangzhi chubanshe chuban, 1996), 436–39.

12. Averill, "Revolution in the Highlands," 23–26, 30.

13. Chang Kuo-t'ao, *The Rise of the Chinese Communist Party, 1921–1927: The Autobiography of Chang Kuo-t'ao* (Lawrence: University of Kansas Press, 1971), 23.

14. Xi Rong, *Pingxiang xianzhi* (Taibei: Ch'engwen chubanshe, 1975), 119; Burton Pasternak, *Kinship and Community in Two Chinese Villages* (Stanford: Stanford University Press, 1972), 18–19; Averill, "Revolution in the Highlands," 36.

15. James M. Polachek, "The Moral Economy of the Kiangsi Soviet (1928–1934)," *Journal of Asian Studies* 42, 4 (1983): 810. For more on this, see Hill Gates, *China's Motor: A Thousand Years of Petty Capitalism* (Ithaca: Cornell University Press, 1996), 84–102.

16. In his study of lineages in the New Territories, for example, Faure shows that lineages provided the membership with access to land, protection against feuding neighbors, ad hoc justice systems, connections to the formal government, and rights to a safety net that protected the family members. See David Fauer, *The Structure of Chinese Rural Society: Lineage and Village in the Eastern New Territories, Hong Kong* (Hong Kong: Oxford University Press, 1986), 36–44, 129–40.

17. Xi, *Pingxiang xianzhi*, 119, quoted in Averill, "Revolution in the Highlands," 37.

18. Gu Jiaxiang, *Chouban Pingxiang tielu gong du* (Pingxiang: Pingxiang xian chubanshe, 1900), 3/5a–7b, 4/23a. See also, Philip C. C. Huang, *The Peasant Family and Rural Development in the Yangzi Delta, 1350–1988* (Stanford: Stanford University Press, 1990), 120.

19. For a description of degree-holders and lineage power in northern China, see Philip C. C. Huang, *The Peasant Economy and Social Change in North China* (Stanford: Stanford University Press, 1985), 231–37.

20. For comparison in his study of Zhejiang Province, for instance, Schoppa wrote that one of the most peripheral counties contained only two men who achieved the *jinshi* degree during the Qing era, while one of the core counties saw twelve men achieve the *jinshi* and fully eighty *juren* scholars in just the Guangxu reign period. See R. Keith Schoppa, *Chinese Elites and Political Change: Zhejiang Province in the Early Twentieth Century* (Cambridge: Cambridge University Press, 1982), 52.

21. Timothy Brook, "Family Continuity and Cultural Hegemony: The Gentry of Ningbo, 1368–1911," in *Chinese Local Elites and Patterns of Dominance*, ed. Joseph W. Esherick and Mary Backus Rankin (Berkeley: University of California Press, 1990), 30.

22. Schoppa, *Chinese Elites*, 52–54.

23. See, for example, a discussion of the son of the powerful gentry member Li Youfen in Liu, *Zhaoping zhilue*, 1474–78.

24. Ch'u Tung-tsu, *Local Government in China under the Ch'ing* (Stanford: Stanford University Press, 1962); Bradly W. Reed, *Talons and Teeth: County Clerks and Runners in the Qing Dynasty* (Stanford: Stanford University Press, 2000).

25. Liu, *Zhaoping zhilue*, 855–58, 862–67.

26. For a more complete discussion of this relationship between landlord and peasant, see James Scott, *Moral Economy of the Peasant: Rebellion and Subsistence in Southeast Asia* (New Haven: Yale University Press, 1976), 13–34.

27. Chang, *Rise of the Chinese Communist Party*, 23.

28. A picul of grain is calculated at various conversion weights. Huang and Wakeman place the weight as 133.33 pounds. See, especially, Huang, *Peasant Family*, 41, 122, and Frederick Wakeman Jr., *The Great Enterprise: The Manchu Reconstruction of Imperial Order in Seventeenth-Century China* (Berkeley: University of California Press, 1985), xiii.

29. Chang, *Rise of the Chinese Communist Party*, 11.

30. Kathryn Bernhardt makes this case for families in the Yangzi delta area during the Qing era. See in particular Kathryn Bernhardt, *Rents, Taxes, and Peasant Resistance: The Lower Yangzi Region, 1840–1950* (Stanford: Stanford University Press, 1992), 15–27. For a description of gentry-dominated neighborhoods in the cities, see Chang Sen-dou, "Urban Geography of the Chinese Hsien Capital," *Annals of the Association of American Geographers* 51, 1 (1951): 23–45. In an interview I had with a member of the powerful Wen lineage, he noted that they lived in the city not far from the county yamen and the Confucian Temple.

31. To be sure, these numbers should all be taken as suggestive at best. Since they are the best figures we have, I present them here but accept their problematic basis.

32. Joseph W. Esherick determines that 39 percent of all Jiangxi provincial households were tenant households and that 34 percent of all households were owner-tenant households. Furthermore, he argues that 59 percent of all land was rented. See his article "Number Games: A Note on Land Distribution in Prerevolutionary China," *Modern China* 7, 4 (1981): 395–97. Similarly, Arrigo finds that in the "Rice-Tea" area, the section of the country that includes Pingxiang County, on average 45.7 percent of the land was rented and that the richest 10 percent of all farmers owned 14.6 percent of all the land. See Linda Gale Arrigo, "Landownership Concentration in China: The Buck Survey Revisited," *Modern China* 12, 3 (1986): 352, 355. See also Joseph W. Esherick, *Reform and Revolution in China: The 1911 Revolution in Hunan and Hubei* (Berkeley: University of California Press, 1976), 59–60; and Mao Zedong, "Report on an Investigation of the Peasant Movement in Hunan," in *Selected Works of Mao Tse-tung*, Vol. 1 (Beijing: Foreign Languages Press, 1967), 30–32.

33. Roy Hofheinz Jr., "Ecology of Chinese Communist Success: Rural Influence Patterns, 1923–45," in *Chinese Communist Politics in Action*, ed. A. Doak Barnett (Seattle: University of Washington Press, 1969), 60.

34. Mao Zedong, *Report from Xunwu*, trans. with an intro. by Roger R. Thompson (Stanford: Stanford University Press, 1990), 161–67.

35. Mao, *Report from Xunwu*, 167.

36. Liu, *Zhaoping zhilue*, 2368–69.

37. See, for example, Huang, *Peasant Family*, and Bernhardt, *Rents, Taxes, and Peasant Resistance*.

38. Averill, "Revolution in the Highlands," 49; Osborne, "Barren Mountains, Raging Rivers," 116–19, 141.

39. Bernhardt, *Rents, Taxes, and Peasant Resistance*, 21–27; Isett, *State, Peasant, and Merchant*, 99–105; Robert Brenner and Christopher Isett, "England's Divergence from China's Yangzi Delta: Property Relations, Microeconomics, and Patterns of Development," *Journal of Asian Studies* 61, 2 (2002): 614–17.

40. The earliest Chinese contracts dating to the Zhou era do not contain this term. See Terry F. Kleeman, "Land Contracts and Related Documents," in *Chugoku no shukyo shiso to kagaku: Makio Ryokai Hakushi shojo kinen ronshu*, ed. Makio Ryokai (Tokyo: University of Tokyo Press, 1986). From the Tang/Song

centuries the idea was almost universal that selling one's ancestral property was only to be a short-term plan to acquire money until the person was able to regain sufficient funds to buy the land back. See Valerie Hansen, *Negotiating Daily Life in Traditional China: How Ordinary People Used Contracts, 600–1400* (New Haven: Yale University Press, 1995), 24–33.

41. Huang, *Peasant Family*, 106.

42. Huang, *Peasant Family*, 106; George Jamieson, *Chinese Family and Commercial Law* (Hong Kong: Vetch and Lee, 1970), 88; Isett, *State, Peasant, and Merchant*, 85.

43. Kang Chao, "New Data on Land Ownership Patterns in Ming-Ch'ing China: A Research Note," *Journal of Asian Studies* 40, 4 (1981): 732–33.

44. Huang, *Pingxiang shi zhi*, 403.

45. See the analysis of land contracts in chapter 4.

46. Pingxiang mei ju, "Pingxiang meikuang chanye qi" (Unpublished), "Table of Contents."

47. Liu, *Zhaoping zhilue*, 1945.

48. County Magistrate Gu's estimates would lead one to believe that every acre of land produced over 8,600 pounds of grain or that every hectare produced 10,189 kilos.

49. *Wen jiapu* (unpublished, undated), 3/3/5b.

50. The best estimates are that on average one hectare probably produced only 2,178.8 kilos of grain. See Arrigo, "Landownership Concentration," 320. Similarly, in his book on the Yangzi delta, Huang found that the average *mu* of irrigated cropland took in about 400 catties or slightly less than 3 piculs (*shi*) of grain. It is highly unlikely that fields in the Jiangxi provincial highlands could outperform the paddy fields of the Jiangnan basin.

51. In her study, Osborne showed that mountain properties in the eastern Jiangxi highland counties were under the control of lineages throughout the Qing era, at least a century before Pingxiang County in the west. See Osborne, "Barren Mountains, Raging Rivers," 113–21.

52. Liu, *Zhaoping zhilue*, 900.

53. Hua Wen and Luo Xiao, "Pingxiang meitan fazhan," in *Pingxiang meitan fazhan shilue*, ed. Jiangxi sheng zheng xi wenshi ziliao yanjiu weiyuan hui and Pingxiang shi zheng wei wenshi ziliao yanjiu weiyuan hui he bian (Hong Kong: N.p., 1987), 2.

54. Luo Xiao, *Pingxiang shi difang meitan gongye zhi* (Nanchang shi: Jiangxi renmin chubanshe, 1992), 1.

55. For further discussion of lineage trusts, see Madeline Zelin, "The Rise and Fall of the Fu-Rong Salt-Yard Elite: Merchant Dominance in Late Qing China," in *Chinese Local Elites and Patterns of Dominance*, ed. Joseph W. Esherick and Mary Backus Rankin (Berkeley: University of California Press, 1990); and Kenneth Pomeranz, " 'Traditional' Chinese Business Firms Revisited: Family, Firm, and Financing in the History of the Yutang Company of Jining, 1779–1956," *Late Imperial China* 18, 1 (1997): 1–38. Gates also describes this phenomenon,

terming the kin-based firms "Patricorporations," showing that the kinship system was both a creation of and a response to the power of the state. See Gates, *China's Motor*, 103–20.

56. Pomeranz, " 'Traditional' Chinese Business Firms"; Fauer, *Structure of Chinese Rural Society*.

57. Averill, "Revolution in the Highlands," 20. In his posthumously published book, Averill defines "mountain lords" as "those who claimed possession of hill lands" and further notes that they "worked on their own or through intermediate hiring agents with connections in the areas of guest people origin to contract for the services of large numbers of guest people." See Stephen Averill, *Revolution in the Highlands: China's Jinggangshan Base Area* (Lanham: Rowman & Littlefield, 2006), 26.

58. Luo, *Pingxiang shi difang*, 1. Since none of the founders appear to have been degree-holders, the local gazetteer provides us with no more information about these men.

59. Luo, *Pingxiang shi difang*, 47.

60. Liu, *Zhaoping zhilue*, 1550–51. The Western technology discussed in the quote was likely dynamite or powered winches and indicates a greater level of capital than was invested by most of the other founders.

61. Liu, *Zhaoping zhilue*, 1179; Hua and Luo, "Pingxiang meitan fazhan," 2.

62. Hua and Luo, "Pingxiang meitan fazhan," 2

63. Chang, *Rise of the Chinese Communist Party*, 5.

64. Hsiao Kung-chuan, *Rural China: Imperial Control in the Nineteenth Century* (Seattle: University of Washington Press, 1960), 419–21.

65. As will be obvious in later chapters, these sites were to be the centerpieces of the mechanized coalmines of the county. Even as the Wen lineage emerged as the most important of the lineages in the early months and years of the modernization process, the mining town where the Western machinery was located was called Anyuan, and the General Horizontal Adit, the main entryway of the mines, was located in Tianzi Mountain.

66. Luo, *Pingxiang shi defang*, 47–48.

67. Wen jiapu, 3/3/12a–12b.

68. Hua and Luo, "Pingxiang meitan fazhan," 6.

69. Sun shows that some mines contracted professional mine laborers to work year-round. See Sun, "Mining Labor, 54."

70. Tim Wright, "Method of Evading Management: Contract Labor in Chinese Coal Mines before 1937," *Comparative Studies in Society and History* 23, 4 (1981): 663.

71. Sun, "Mining Labor," 53–54.

72. Wright, "Method of Evading Management," 324–27; Leonard G. Ting, "Coal Industry in China," Part I, *Nankai Social and Economic Quarterly* 10, 1 (1937): 73.

73. "Investigation Report," in *Chinese Civilization and Society: A Sourcebook*, ed. Patricia Ebrey (New York: Free Press, 1981), 233. See also Sun, "Mining Labor," 56; and Ting, "Coal Industry of China," 73.

74. Ting, "Coal Industry of China," 250.

75. Luo, *Pingxiang shi difang*, 48; Liu Minghan and Ma Jingyuan, eds., *Hanyeping gongsi zhi* (Wuhan: Huazhong li gong daxue chubanshe, 1990), 60–62.

76. Chen Xulu and Gu Tinglong, eds., *Hanyeping gongsi*, Vol. 1 (Shanghai: Shanghai renmin chubanshe, 1984), 136; Luo, *Pingxiang shi difang*, 47; Averill, "Revolution in the Highlands," 20; Contracts: "Table of Contents." Note that throughout this book when I describe indigenous mines, I call them "coalmines," but when I am describing the mines under the control of the government's Mine Bureau, I capitalize the term and refer to the mechanized mines as the "Pingxiang County Coalmines" or simply the "Coalmines."

77. Chen and Gu, *Hanyeping gongsi*, Vol. 1, 115–16.

78. Luo, *Pingxiang shi difang*, 48.

79. Chen and Gu, *Hanyeping gongsi*, Vol. 1, 109–110, 196.

80. This intensification of output without a qualitative change in production schemes is what Huang refers to in the farming economy in China as "the paradoxical concurrence of vigorous commercialization and subsistence-level farming." See Philip C. C. Huang, "The Paradigmatic Crisis in Chinese Studies: Paradoxes in Social and Economic History," *Modern China* 17, 3 (1991): 299–341.

81. James Scott, *Weapons of the Weak: Everyday Forms of Peasant Resistance* (New Haven: Yale University Press, 1985), and James Scott, *Domination and the Arts of Resistance: Hidden Transcripts* (New Haven: Yale University Press, 1990).

82. Liu, *Zhaoping zhilue*, 1127–29, 1692–93; Philip A. Kuhn, *Rebellion and Its Enemies in Late Imperial China: Militarization and Social Structure* (Cambridge: Harvard University Press, 1980), 200–02; Averill, "Revolution in the Highlands," 69, 72.

83. Kuhn, *Rebellion and Its Enemies*. See also Hsiao, *Rural China*, 294–306.

84. Liu, *Zhaoping zhilue*, 1131–33; Esherick, *Reform and Revolution*, 21–24; Fei-ling Davis, *Primitive Revolutionaries of China: A Study of Secret Societies of the Late Nineteenth Century* (Honolulu: University of Hawaii Press, 1971), 91–92.

85. Davis, *Primitive Revolutionaries*, 86, see also 130–36.

86. G. William Skinner, "Marketing and Social Structure in Rural China," Part I, *Journal of Asian Studies* (1965): 37; Davis, *Primitive Revolutionaries*, 79–80.

87. Esherick, *Reform and Revolution*, 59.

88. Chang, *Rise of the Chinese Communist Party*, 4, 6; Charlton M. Lewis, *Prologue to the Chinese Revolution: The Transformation of Ideas and Institutions in Hunan Province, 1891–1907* (Cambridge: East Asian Research Center, Harvard University Press, 1976), 78.

89. Samuel Yale Kupper, "Revolution in China: Kiangsi Province, 1905–1913," PhD dissertation (University of Michigan, 1973), 80.

90. Davis, *Primitive Revolutionaries*, 176.

91. For similar discussions of popular support for illicit organizations, see Eric Hobsbawm, *Bandits*, Revised Edition (New York: Random House, 1981), 17–29.

92. Davis, *Primitive Revolutionaries*, 72–79, 109, 151; Liu, *Zhaoping zhilue*, 1131–33.

93. Polachek, "Moral Economy," 812.

94. Gail Hershatter, *The Workers of Tianjin, 1900–1949* (Stanford: Stanford University Press, 1986), 197; Kimio Nishizawa, "Mines in the Yangtze Valley, China," *Journal of the Royal Society of Arts* 58 (1910): 944.

95. Hershatter, *Workers of Tianjin*, 202.

96. Liu Kunyi, *Liu Kunyi yiji* (Beijing: Zhonghua shuju chubanshe, 1959), 232.

97. Liu, *Zhaoping zhilue*, 1129–30.

98. Esherick, *Reform and Revolution*, 59.

99. Phil Billingsley, "Bandits, Bosses, and Bare Sticks: Beneath the Surface of Local Control in Early Republican China," *Modern China* 7, 3 (1981): 235–88.

Chapter 3. Self-Strengthening Up Above and Reorganizing Down Below

1. See Alfred D. Chandler, *The Visible Hand: The Managerial Revolution in American Business* (Cambridge: Harvard University Press, 1977), and Alfred D. Chandler, "The Emergence of Managerial Capitalism," *Business History Review* 58, 4 (1984): 473–503; and Olivier Zunz, *Making America Corporate, 1870–1920* (Chicago: University of Chicago Press, 1990).

2. Elisabeth Köll, *From Cotton Mill to Business Empire: The Emergence of Regional Enterprises in Modern China* (Cambridge: Harvard University Press, 2003), 22; Philip C. C. Huang, *The Peasant Family and Rural Development in the Yangzi Delta, 1350–1988* (Stanford: Stanford University Press, 1990).

3. Elisabeth Köll, "Recent Debates in the Field of Business History: What They Mean for China Historians," *Chinese Business History* 10, 1 (2000): 1–2.

4. Albert Feuerwerker, *China's Early Industrialization: Sheng Hsuan-huai (1844–1916) and Mandarin Enterprise* (New York: Atheneum, 1970).

5. Wu Chang-chuan, "Cheng Kuang-ying: A Case of Merchant Participation in the Chinese Self-Strengthening Movement (1878–1884)," PhD dissertation (Columbia University, 1974); Yuen-sang Leung, *The Shanghai Taotai: Linkage Man in the Changing Society, 1843–90* (Honolulu: Asian Studies at Hawaii, 1990).

6. "Take-off," as it was described by W. W. Rostow, was essentially brought about by private investments and advancements. However, Gerschenkron argues that in the less-developed countries and empires like Russia in the late nineteenth century, active state investment led to modern growth. See Alexander Gerschenkron, *Economic Backwardness in Historical Perspective: A Book of Essays* (Cambridge: Harvard University Press, 1962).

7. Ray Huang, "Lung-ch'ing and Wan-li Reigns, 1567–1620," in *The Cambridge History of China: Volume 7: The Ming Dynasty, 1368–1644, Part I*, ed. Frederick W. Mote and Denis Twitchett (Cambridge: Cambridge University Press, 1988), 530–32; Shih-shan Henry Tsai, *Eunuchs in the Ming Dynasty* (Albany: State University of New York Press, 1996), 111, 182.

8. Fang Zhoufen, Hu Tiewen, Juan Rui, and Fang Xing, "Capitalism during the Early and Middle Qing," in *Chinese Capitalism, 1522–1840*, ed. Xu Dixin and Wu Chengming, ed. and annotated by C. A. Curwen, trans. Li Zhengse, Liang Mioru, and Li Siping (New York: St. Martin's Press, 2000), 249–53, 289–96.

9. Sun E-tu Zen, "Ch'ing Government and the Mineral Industries before 1800," *Journal of Asian Studies* 27, 4 (1968): 835–45; H. C. Hoover, "Present Situation of the Mining of Industry of China," *Engineering and Mining Journal* 69 (1900): 619.

10. Peter J. Golas, *Science and Civilization in China: Volume 5: Chemistry and Chemical Technology: Part XII: Mining* (Cambridge: Cambridge University Press, 1999), 416–28; see also Chen fu-mei Chang and Ramon Myers, "Customary Law and Economic Growth of China During the Ch'ing Period," Part II, *Ch'ing-shih wen-t'i* 3, 10 (1976): 4–27; Liu Hongpi, ed., *Zhaoping zhilue* (Taiwan: Chengwen chubanshe, 1975), 892–93. For a contemporary description of government "ambivalence," see Albert S. Bickmore, "Sketch of a Journey from Canton to Hankow," *Journal of the Royal Geographical Society of London* 38 (1868): 67–68.

11. For a discussion of regional government and industrialization, see Daniel H. Bays, "The Nature of Provincial Political Authority in Late Ch'ing Times," *Modern Asian Studies* 4, 4 (1970): 325–47; and Kwang-Ching Liu, "The Beginnings of Modernization," in *Li Hung-chang and China's Early Modernization*, ed. Samuel C. Chu and Kwang-Ching Liu (Armonk: M. E. Sharpe, 1994). Landes argues that even after seeing the power of the West many Chinese leaders continued with their belief in the inherent superiority of the Chinese ideal. His argument is no doubt true for many Chinese officials. However, the actions of many regional political leaders suggests that they had been converted, even if begrudgingly, to the belief that incorporating Western technology into their society was ironically required to save it. See David S. Landes, T*he Unbound Prometheus: Technological Change and Industrial Development in Western Europe from 1750 to the Present* (Cambridge: Cambridge University Press, 1969), 28. See also Luke S. K. Kwong, "Ti-Yung Dichotomy and the Search for Talent in Late-Ch'ing China," *Modern Asian Studies* 27, 2 (1993): 253–79.

12. Feuerwerker, *China's Early Industrialization*, 106; Ellsworth C. Carlson, *The Kaiping Mines 1877–1912* (Cambridge: Harvard University Press, 1971), 7, 43; Tim Wright, *Coal Mining in China's Economy and Society, 1895–1937* (Cambridge: Cambridge University Press, 1984), 141, 144; Liu, "The Beginnings of Modernization," 10; Leonard G. Ting, "The Coal Industry of China," Part I, *Nankai Social and Economic Quarterly* 10, 1 (1937): 41–43.

13. Chi-kong Lai, "Li Hung-chang and Modern Enterprise: The China Merchants' Company, 1872–1885," in *Li Hung-chang and China's Early Modernization*, ed. Samuel C. Chu and Kwang-China Liu (Armonk: M. E. Sharpe, 1994), 221.

14. William Reid Johnson, "Hanyang Iron Works 1890–1908: A Key Enterprise in China's Industrialization," MA thesis (University of Washington, 1955), 39; Wu, "Cheng Kuan-ying," 254.

15. Feuerwerker, *China's Early Industrialization*, 170–72; Carlson, *Kaiping Mines*, 8, 44–45; Hoover, "Present Situation of the Mining Industry of China," 619.

16. Seungjoo Yoon, "Constitutional Change in the Lower Echelon of the Qing Bureaucracy: The Formation, Reformation, and Transformation of Zhang Zhidong's Document Commissioners, 1885–1909," PhD dissertation (Harvard University, 1999), 63–66; En-han Lee, "China's Response to Foreign Investment in Her Mining Industry," *Journal of Asian Studies* 28, 1 (1968): 59.

17. Feuerwerker, *China's Early Industrialization*, 115–16.

18. Thomas L. Kennedy, "Chang Chih-tung and the Struggle for Strategic Industrialization: The Establishment of the Hanyang Arsenal, 1884–1895," *Harvard Journal of Asiatic Studies* 30 (1973): 157–58; Bays, "The Nature of Provincial Political Authority," 334–35.

19. Kwong, "The Ti-Yung Dichotomy."

20. Ssu-yu Teng and John K. Fairbank, eds., *China's Response to the West: A Documentary Survey, 1839–1923* (New York: Atheneum, 1970), 171.

21. Philip Yuen-sang Leung, "Crisis Management and Industrial Reform: The Expectant Officials in the Late Qing," in *Dragons, Tigers, and Dogs: Qing Crisis Management and the Boundaries of State Power in Late Imperial China*, ed. Robert J. Anthony and Jane Kate Leonard (Ithaca: Cornell University Press, 2002).

22. Yoon, "Constitutional Change," 63. In his study, Yoon lays out Zhang's organization showing both the levels of power and of specialization, referring to most of these posts as "Document Commissioners," a term I believe confuses some of the different tasks the men were hired to accomplish.

23. Leung, "Crisis Management and Institutional Reform," 64.

24. Bays, "The Nature of Provincial Political Authority," 337; Kennedy, "Chang Chih-tung and the Struggle for Strategic Industrialization," 160; Albert Feuerwerker, "China's Nineteenth-Century Industrialization: The Case of the Hanyeping Coal and Iron Company, Limited," in *The Economic Development of China and Japan: Studies in Economic History and Political Economy*, ed. C. D. Cowan (New York: Frederick A. Praeger, 1964), 87; F. R. Tegengren, *Iron Ores and Iron Industry of China, Part II* (Beijing: Geological Survey of China, Ministry of Agriculture and Commerce, 1923–1924), 366.

25. Tegengren, *Iron Ores and Iron Industry of China*, 366.

26. Johnson, "Hanyang Iron Works," 64; Collins, *Mineral Enterprise in China*, 78.

27. Orodic Y. K. Wou, "Development, Underdevelopment and Degeneration: The Introduction of Rail Transport into Honan," *Asian Profile* 7, 3 (1984): 217.

28. For a map of the railroad line from Beijing to Wuhan, see Thomas T. Read, "The Mineral Production and Resources of China," *Transactions of the American Institute of Mining Engineers* 43 (1912): 298.

29. Kennedy, "Chang Chih-tung and the Struggle for Strategic Industrialization," 160.

30. Tegengren, *Iron Ores and Iron Industry of China*, 366.

31. Kennedy, "Chang Chih-tung and the Struggle for Strategic Industrialization," 163.

32. "Notes on Coal Mines in China," *The Far Eastern Review* 37 (1931): 203; K. L. Hsueh, "The Iron and Steel Industry of China," *Chinese Economic Journal* 2, 1 (1928): 1. (This article was reprinted as K. L. Hsueh, "The Iron and Steel Industry of China," *The Far Eastern Review* 24, 4 (1928): 176–83.)

33. Yoon, "Constitutional Change," 63–64.

34. Johnson, "Hanyang Iron Works," 85–87; Lansing Hoyt, "Blast Furnaces and Steel Mills in China: A Comparative Study of China's Steel Industry," *The Far Eastern Review* 19, 5 (1923): 307; Percy M. Roxby, "Wu-Han: The Heart of China," *Scottish Geographical Magazine* 32 (1916): 275–76; Tegengren, *Iron Ores and Iron Industry of China*, 373; Zhongguo shi xuehui, Zhongguo ke xueyuan, and Zongguo yang danganguan, eds., *Yangwu Yundong*, Vol. 7 (Shanghai: Shanghai renmin chubanshe and Shanghai shu dian chubanshe, 2000), 297–98. [Henceforth referred to as *YWYD*].

35. Chen Xulu and Gu Tinglong, eds., *Hanyeping gongsi*, Vol. 1 (Shanghai: Shanghai renmin chubanshe, 1984), 10.

36. Johnson, "Hanyang Iron Works," 76.

37. Chen and Gu, *Hanyeping gongsi*, Vol. 1, 9.

38. Chen and Gu, *Hanyeping gongsi*, Vol. 1, 28–29, 40, 54, 79.

39. Chen and Gu, *Hanyeping gongsi*, Vol. 1, 58–60.

40. Liu Ping, *Zhang Zhidong zhuan* (Lanzhou: Lanzhou daxue chubanshe, 2000), 196–97; Hsueh, "Iron and Steel Industry of China," 19–20.

41. Gu Lang, *Zhongguo shi da kuang chang diaocha ji* (Shanghai: Commercial Press, 1916), chap. 3, pp. 3, 13; Wright, *Coal Mining in China's Economy and Society*, 44.

42. Liu, *Zhaoping zhilue*, 78, 1232.

43. Yoon, "Constitutional Change," 170.

44. Chen Zhen, *Zhongguo jindai gongye shi ziliao* (Beijing: N.p., 1961), 443–44; Hua Wen and Luo Xiao, "Pingxiang meitan fazhan shi lue," in *Pingxiang meitan fazhan shilue*, ed. Jiangxi sheng zheng xi wenshi ziliao weiyuan hui and Pingxiang shi zheng wen weshi ziliao yanjiu weiyuan hui he bian (Hong Kong: N.p., 1987), 6.

45. Chen and Gu, *Hanyeping gongsi*, Vol. 1, 84; Luo Xiao, *Pingxiang shi difang meitain gong ye zhi* (Documents of Pingxiang City's Local Coal Production) (Nanchang shi: Jiangxi renmin chubanshe, 1992), 47–49.

46. Chen and Gu, *Hanyeping gongsi*, Vol. 1, 28–29.

47. Feuerwerker, "China's Nineteenth-Century Industrialization," 87; see also, Kennedy, "Chang Chih-tung and the Struggle for Strategic Industrialization," 175.

48. Feuerwerker, *China's Early Industrialization*, 60–61.

49. Yoon, "Constitutional Change," 78.

50. Chen and Gu, *Hanyeping gongsi*, Vol. 1, 23–26, 27–28, 45–47, 55–57.

51. Chen and Gu, *Hanyeping gongsi*, Vol. 1, 40–42.

52. Feuerwerker, "China's Nineteenth-Century Industrialization," 87.

53. Feuerwerker, *China's Early Industrialization*, 67–68; E. T. Williams, *Recent Chinese Legislation Relating to Commercial, Railway, and Mining Enterprises with Regulations for Registration of Trade Marks, and for the Registration of Companies* (Shanghai: Shanghai Mercury, 1905), 4.

54. Chen and Gu, *Hanyeping gongsi*, Vol. 1, 40–42, 58–64.

55. Chen and Gu, *Hanyeping gongsi*, Vol. 1, 71; Yoon, "Constitutional Change," 81.

56. *YWYD*, Vol. 7, pp. 268–69.

57. Quoted in Feuerwerker, "China's Nineteenth-Century Industrialization," 99.

58. Luo, *Pingxiang shi difang gongye zhi*, 48. For an example of how the Chinese formed a similar coalmining partnership, see Chen and Myers, "Customary Law and the Economic Growth of China," 20. For more on Sheng Xuanhuai's relationship with the Wen lineage, see Jeff Hornibrook, "Riding the Tiger: Merchant-State Alliance in a Coalmine Modernisation Scheme," *Business History* 45, 2 (2003): 35–51.

59. Chen and Gu, *Hanyeping gongsi*, Vol. 1, 115; Luo, *Pingxiang shi difang meitan*, 48.

60. Reed, "Coal Mining in China," 45–47.

61. Chen and Gu, *Hanyeping gongsi*, Vol. I, 137, 196, 230.

62. Chen and Gu, *Hanyeping gongsi*, Vol. I, 115.

63. Chen and Gu, *Hanyeping gongsi*, Vol. I, 194.

64. Arthur W. Hummel, ed., *Eminent Chinese of the Ch'ing Period (1644–1912)* (Washington: United States Printing Office, 1943), 855–56; Liu, *Zhaoping zhilue*, 1166, 1462; Huang Shigao, ed., *Pingxiang shi zhi* (Pingxiang: Pingxiang shi zhi bianzuan weiyuan huibian gangzhi chubanshe chuban, 1996), 1184–85. Wen Tingshi was a prolific writer who was interested in many issues of modernization and education. However, I have not found any writings by him on the matters discussed in this book. For both a brief history of Wen Tingshi and his family as well as a discussion of his activities in the years of this study, see Wen Tingshi, *Wen Yunge xiansheng quanji* (Taipei: Wenhai chubanshe, 1975), juan 1, sec. 4, 1–2, 31–57. For more of my research on the Wen lineage and the coalmines in Pingxiang County, see Jeff Hornibrook, "Local Elites and Mechanized Mining in China," *Modern China* 27, 2 (2001): 202–28.

65. Chen and Gu, *Hanyeping gongsi*, Vol. 1, 260.

66. Chen and Gu, *Hanyeping gongsi*, Vol. 1, 114, 222.

67. Chen and Gu, *Hanyeping gongsi*, Vol. 1, 259–60.

68. Chen and Gu, *Hanyeping gongsi*, Vol. 1, 115.

69. Chen and Gu, *Hanyeping gongsi*, Vol. 1, 110, 136.

70. Chen and Gu, *Hanyeping gongsi*, Vol. 1, 163–66. Conversely, Susan Mann writes that *lijin bureaux* in Hunan Province were uniquely staffed with both gentry and clerks to reduce corruption. See Susan Mann, *Local Merchants and the Chinese Bureaucracy, 1750–1950* (Stanford: Stanford University Press, 1987), 110–11.

71. Chen and Gu, *Hanyeping gongsi*, Vol. 1, 110, 136.

72. Chen and Gu, *Hanyeping gongsi*, Vol. 1, 222.

73. Chen and Gu, *Hanyeping gongsi*, Vol. 1, 446–47.

74. Chen and Gu, *Hanyeping gongsi*, Vol. 1, 317–18, 437–39, 494; Lu, *Zhaoping zhilue*, 78, 1232.

75. Liu, *Zhaoping zhilue*, 51–52; Fu Xiongxiang, ed., *Liling xiangtu zhi* (Taiwan: Chengwen chubanshe, 1926), 24–26; Huang Zuxun, ed., *Liuyang xiangtu zhi* (Taibei: Qingshui yingshua chubanshe, 1967), 251.

76. Chen and Gu, *Hanyeping gongsi*, Vol. 1, 436–37.

77. Luo, *Pingxiang shi difang*, 49.

78. Chen and Gu, *Hanyeping gongsi*, Vol. 1, 438.

79. Chen and Gu, *Hanyeping gongsi*, Vol. 1, 432–40.

80. This charge is, of course, blatantly false since Lu Hongchang had already met with Wen Tingshi, visited the Guangtaifu mines, and was well aware of the political power the Wen lineage held in the county.

81. Chen and Gu, *Hanyeping gongsi*, Vol. 1, 446–47.

82. Liu, *Zhaoping zhilue*, 1141–1240, 1464–66, 1535.

83. The petition was a means by which commercial leaders could inform the local government of their frustrations with policies they deemed restrictive or unjust. See Richard John Lufrano, *Honorable Merchants: Commerce and Self-Cultivation in Late Imperial China* (Honolulu: University of Hawaii Press, 1997), 80.

84. Chen and Gu, *Hanyeping gongsi*, Vol. 1, 494.

85. Chen and Gu, *Hanyeping gongsi*, Vol. 1, 606–09.

86. Gu, *Zhongguo shi da kuang chang*, chap. 3, p. 5.

87. Feuerwerker, *China's Early Industrialization*, 88; Sheng Xuanhuai, *Yujai congao* (Taibei: Wenhai chubanshe, 1974), 367–69; Ping Wen, "Zhang Zanchen jianjie," in *Pingxiang meitan fazhan shilue*, ed. Jiangxi sheng zheng xi wenshi ziliao weiyuan hui and Pingxiang shi zheng wen weshi ziliao yanjiu weiyuan hui he bian (Hong Kong: N.p., 1987), 194.

88. Chen and Gu, *Hanyeping gongsi*, Vol. 1, 315–18, 337–38. Two Western articles written about the mines in Pingxiang County—one written in 1916 and the other in 1917—provide a photo of Zhang. The first article used his literary name spelled in Wade-Giles and the words "Chang Shou-chan, the Company's First Manager." The second article similarly simply writes "Mr. Chang: Chinese Founder of the Ping Hsiang Colliery." See Alfred C. Reed, "Coal Mining in China," *The Scientific Monthly* 5 (1917): 49, and "The Pingsiang Colliery: A Story of Early Mining Difficulties in China," *The Far Eastern Review* 12, 10 (1916): 375. In fact, nearly every history of the Pingxiang County Coalmines establishes 1898 as the year of the founding of the mines based on Zhang Zanchen's ascension to this office.

89. Gu, *Zhongguo shi da kuang chang*, chap. 3, pp. 5–6.

90. "Pingsiang," 379.

91. Gu, *Zhongguo shi da kuang chang*, chap. 3, pp. 5–6.

92. Chen and Gu, *Hanyeping gongsi*, Vol. 1, 632–33; "Pingsiang," 380.

93. Chen, *Zhongguo jindai gongye shi ziliao*, 444; Gu, *Zhonguo shi da kuang chang*, chap. 3, p. 8.

94. Chen and Gu, *Hanyeping gongsi*, Vol. 1, 632–33.

95. Chen and Gu, *Hanyeping gongsi*, Vol. 1, 116; Elizabeth J. Perry, *Anyuan: Mining China's Revolutionary Tradition* (Berkeley: University of California Press, 2012), 17–19.

96. Chen, *Zhongguo jindai gongye shi ziliao*, 441–43; Feuerwerker, "China's Nineteenth Century Industrialization," 89; "Pingsiang," 375.

97. William Barclay Parsons, *Railways in China* (Philadelphia: Engineers Club of Philadelphia, 1916), 50, 60; "Pingsiang," 375. In some cases Western banks demanded that engineering be done by someone hired from their own country. It is not certain if that requirement was part of this loan or not.

98. Mauris B. Jansen, "Yawata, Hanyehping, and the Twenty-One Demands," *Pacific Historical Review* 23, 1 (1954): 32–33.

99. Chen and Gu, *Hanyeping gongsi*, Vol. 1, 84–85.

100. *YWYD*, Vol. 7, 268–69.

101. Chen and Gu, *Hanyeping gongsi*, Vol. 1, 228–29.

102. "Pingsiang," 377.

103. Zhongguo shixuehui, ed., *Xinhai geming* (Shanghai: Renmin chubanshe, 1957), 480. [Hereafter referred to as *XHGM*.]

104. Chang, *Rise of the Chinese Communist Party*, 16; Chen and Gu, *Hanyeping gongsi*, Vol. 1, 237.

105. Chen and Gu, *Hanyeping gongsi*, Vol. 2, 32–33.

106. Liu, *Zhaoping zhilue*, 896–99.

107. "Pingsiang," 375–76. This description of the coal is similar to English coal. See I. C. F. Statham, *Coal Mining Practice*, Vol. 4 (London: Caxton Publishing Company, 1958), 304.

108. Read, "Mineral Production," 33.

109. Chen and Gu, *Hanyeping gongsi*, Vol. 2, 327–29; Read, "Mineral Production," 34; Hsueh, "Iron and Steel Industry," 14.

110. *Encyclopedia Britannica: A Dictionary of Arts, Sciences, and General Literature*, 9th edition (New York: Charles Scribner's Sons, 1881), 282.

111. Tegengren, *Ores and Iron Industry of China*, 270–72.

112. Liu, *Zhaoping zhilue*, 893–94.

113. Eric R. Wolf, *Peasant Wars of the Twentieth Century* (New York: Harper Torchbooks, 1969), 276–302, and Eric R. Wolf, "On Peasant Rebellions," in *Peasants and Peasant Societies: Selected Readings*, ed. Teodor Shanin (New York: Penguin Books, 1971), 267–69.

Chapter 4. Irrevocably Remapping the County

1. For discussion of the terms "deterritorialization" and "reterritorialization," see James L. Hevia, *English Lessons: The Pedagogy of Imperialism in Nineteenth-Century China* (Durham: Duke University Press, 2003), 21–22, 49–73.

2. Studies published in the journal *Technology and Culture* show that many tools and technological advancements that Westerners brought into non-Western

countries were socially constructed devices essential to production in the country of origin but not necessarily needed in the receiving culture. To this end, in some cases modernization schemes were accomplished through a conglomeration of the modern and premodern. See, for example, Heather J. Hoag and May-Britt Öhman, "Turning Water into Power: Debates over the Development of Tanzania's Rufiji River Basin, 1945–1985," *Technology and Culture* 49, 3 (2008): 624–51. Others have shown, however, that because the colonizers had far more power than the recipients of the new technology the give and take was far from equal. Not only did Western engineers impose a completely new industrial scheme on the population, but these new devices and their logic led to new forms of oppression. See, for example, Allen Isaacman and Chris Sneddon, "Toward a Social and Environmental History of the Building of Cahora Bassa Dam," *Journal of Southern African Studies* 26, 4 (2000): 597–632, and Daniel R. Headrick, *The Tools of Empire: Technology and European Imperialism in the Nineteenth Century* (New York: Oxford University Press, 1981).

3. See the articles in Madeline Zelin, Jonathan K. Ocko, and Robert Gardella, eds., *Contract and Property in Early Modern China* (Stanford: Stanford University Press, 2004). See also Christopher Mills Isett, *State, Peasant, and Merchant in Qing Manchuria, 1644–1862* (Stanford: Stanford University Press, 2007).

4. In such cases when eminent domain is to be enacted in modern Western culture, the state is usually obligated to provide some compensation to the original rightful landholder. Dating back to early English common law, the crown declared that all the lands within the kingdom were the rightful property of the throne but that commoners were allowed to use land in the interests of the economy of the empire. See Polly J. Price, *Property Rights: Rights and Liberties under the Law* (Santa Barbara: ABC-CLIO, 2003), 9.

5. Gu Lang, *Zhonguo shi da kuang chang diaocha ji* (Shanghai: Commercial Press, 1916), chap. 3, pp. 2–6; Liu Hongpi, ed., *Zhaoping zhilue* (Taiwan: Chengwen chubanshe, 1975), 893–94.

6. Chen Xulu and Gu Tinglong, *Hanyeping gongsi*, Vol. 1 (Shanghai: Shanghai renmin chubanshe, 1984), 240–41; Luo Xiao, *Pingxiang shi difang meitang gongye zhi* (Nanchang: Jiangxi renmin chubanshe, 1992), 49; Hua Wen and Luo Xiao, "Pingxiang meitan fazhan shilue," in *Pingxiang meitan fazhan shilue*, ed. Jiangxi sheng zheng xi wenshi ziliao yanjiu weiyuan hui and Pingxiang shi zheng wei wenshi ziliao yanjiu weiyuan hui he bian (Hong Kong: N.p., 1987), 9.

7. Gu, *Zhongguo da shi kuang chang*, chap. 3, pp. 7–8.

8. Gu, *Zhongguo da shi kuang chang*, chap. 3, p. 8; Luo, *Pingxiang shi difang*, 2.

9. Pingxiang kuang wu ju, eds., *Jinian ce: Ping kuang jianshe jiushi zhouhua* (Pingxiang: Ping kuang gongren baoshe yinshuachang, 1988), 7.

10. Elizabeth J. Perry, *Anyuan: Mining China's Revolutionary Tradition* (Berkeley: University of California Press, 2012), 18–19; Huang Shigao, ed., *Pingxiang shi zhi* (Pingxiang: Pingxiang shi zhi bianzuan weiyuan huibian gangzhi chubanshe chuban, 1996), 1185–86.

11. Gu, *Zhongguo shi da kuang chang*, chap. 3, p. 8. I will discuss money and land calculations in more detail later. However, the information here indicates that Zhang Zanchen bought 1,700 *mu* of land with 510,000 taels of silver or an average of 300 taels per *mu*. In fact, the cost may have been significantly less if the 510,000 taels were also used to purchase Guangtaifu as well as other equipment and structures, which is likely.

12. The contracts under discussion here were also briefly examined in chapter 2. For more information on these documents, see Jeff Hornibrook, "Don't Tread on Me: Land, Officials, and Archival Work on a Qing Dynasty Mining Enterprise," *Chinese Business History* 15, 1 (2005): 1–2, 9. In that article I make clear that the contracts I have collected were necessarily haphazard as the archivists were reluctant to assist me in the complete retrieval of these documents. I was given an opportunity to copy about a dozen of the contracts by hand. I then had to walk away from the hundreds of contracts I viewed.

13. Contract #123.

14. Contract #289.

15. Contract #101.

16. Pingxiang kuangwu ju zhi weiyuan hui, eds., *Pingxiang kuangwu ju zhi* (Jiangxi: Pingxiang kuangwu ju zhi bian weiyuan hui, 1988), 62 (hereafter referred to as *PKWJ*); Pingxiang wenshi ziliao (7th division) and Pingxiang shi zhi tongxun (14th edition), eds., *Pingxiang renwu ju lue* (Pingxiang: Pingxiang shi zheng xie wenshi ziliao yanjiu weiyuan hui and Pingxiang shi zhi bian luo weiyuan hui, 1987), 172 (hereafter referred to as *PXRW*).

17. Luo, *Pingxiang shi difang meitan gongye zhi*, 49; Lansing W. Hoyt, "Blast Furnaces and Steel Mills in China: A Comprehensive Study of China's Steel Industry," *The Far Eastern Review* 19, 5 (1923): 315; Gu, *Zhongguo shi da kuang chang*, chap. 3, pp. 8–9.

18. William R. Braisted, "United States and the American China Development Company," *Far Eastern Quarterly* 11, 2 (1952): 147–65.

19. Raymond A. Esthus, "The Changing Concept of the Open Door, 1899–1910," *The Mississippi Valley Historical Review* 46, 3 (1959): 436; En-han Lee, "China's Response to the Foreign Scramble for Railway Concessions, 1895–1911," *Journal of Oriental Studies* 14, 1 (1976): 3; American China Development Company, *Contracts, Chinese Government and American China Development Company: Dated April 14th, 1898, and July 13th, 1900* (China: American China Development Company, 1900–1904) (hereafter referred to as *ACDC*, 1900–1904); William Barclay Parsons, *An American Engineer in China* (New York: McClure, Phillips, and Company, 1900); Charlton M. Lewis, *Prologue to the Chinese Revolution: The Transformation of Ideas and Institutions in Hunan Province, 1891–1907* (Cambridge: East Asian Research Center, Harvard University Press, 1976), 128–31; Chen Xulu and Gu Tinglong, *Hanyeping gongsi*, Vol. 2 (Shanghai: Shanghai renmin chubanshe, 1986), 84–86.

20. William Barclay Parsons, *Report on the Survey and Prospects of a Railway between Hankow and Canton, under the Concession by the Chinese Government to*

the American China Development Company (New York: American China Development Company, 1899), 66; Parsons, *American Engineer in China*, 66.

21. *ACDC*, 36; Parsons, *Report on the Survey and Prospects of a Railway*, 8; Percy Horace Kent, *Railway Enterprise in China: An Account of Its Origin and Development* (London: Edward Arnold, 1907), 121, lists W. W. Rich as "Captain Rich."

22. Gu Jiaxiang, *Chouban Pingxiang tielu gong du* (Pingxiang: Pingxiang xian chubanshe, 1900).

23. Zhang Yi, Zhu Xiangqing, and Fan Yifei, eds., *Jiangxi sheng zhi* (Beijing: Zhong gong zhong yang dang xiao chubanshe, 1993), 9–11; Guo shi guan, eds., *Zhongguo tielu yange shi* (Taibei: Taibei xian xindian shi, 1984), 473.

24. Liu, *Zhaoping zhilue*, 988–89, 1016–18.

25. Zhang et al., *Jiangxi sheng zhi*, 10.

26. Gu, *Chouban Pingxiang tielu gong du*, chap. 1, pp. 8a–8b.

27. Liu, *Zhaoping zhilue*, 1063.

28. Liu, *Zhaoping zhilue*, 988–89, 1016–18; Huang, *Pingxiang shi zhi*, 1183–84.

29. See R. H. Mathews, *Mathews' Chinese-English Dictionary*, Revised American Edition (Cambridge: Harvard University Press, 1931), 8; Liang Shih-shih, ed., *New Practical Chinese-English Dictionary* (Taipei: The Far Eastern Book Company, 1988), 1256.

30. Liu, *Zhaoping zhilue*, 2246–48.

31. Liu, *Zhaoping zhilue*, 58–59.

32. Liu, *Zhaoping zhilue*, 2274–75; Gu, *Chouban Pingxiang tielu gong du*, chap. 3, p. 9a.

33. Gu, *Chouban Pingxiang tielu gong du*, chap. 2, pp. 30b–31a.

34. Gu, *Chouban Pingxiang tielu gong du* chap. 2, pp. 28b–29a; Liu, *Zhaoping zhilue*, 1016–18.

35. Gu, *Chouban Pingxiang tielu gong du*, chap. 2, pp. 30b–31a.

36. Gu, *Chouban Pingxiang tielu gong du*, chap. 3, pp. 7b–8b.

37. Gu, *Chouban Pingxiang tielu gong du*, chap. 3, p. 8b.

38. William Barclay Parsons, *Railways in China* (Philadelphia: Engineers Club of Philadelphia, 1916), 65; Parsons, *American Engineer in China*, 264.

39. Gu, *Chouban Pingxiang tielu gong du*, chap. 3, p. 8b.

40. Gu, *Chouban Pingxiang tielu gong du*, chap. 4, p. 22b.

41. Liu, *Zhaoping zhilue*, 40–41.

42. George Jamieson, *Chinese Family and Commercial Law* (Hong Kong: Vetch and Lee, 1970), 84.

43. Parsons, *American Engineer in China*, 264–65; Thomas T. Read, "The Mineral Production and Resources of China," *Transactions of the American Institute of Mining Engineers* 43 (1912): 294.

44. Parsons, *American Engineer in China*, 265. For a similar complaint, see Geo Bronson Rea, "Railway Loan Agreements and Their Relations to the Open Door: A Plea for Fair Play to China," *The Far Eastern Review* 6, 6 (1909): 221.

45. Hoyt, "Blast Furnaces and Steel Mills," 317.

46. Gu, *Chouban Pingxiang tielu gong du*, chap. 2, p. 28a.

47. Gu, *Chouban Pingxiang tielu gong du*, chap. 3, pp. 6b–7a.

48. Gu, *Chouban Pingxiang tielu gong du*, chap. 3, pp. 5b–7a.

49. Gu, *Chouban Pingxiang tielu gong du*, chap. 3, pp. 5b–7a.

50. Gu, *Chouban Pingxiang tielu gong du*, chap. 4, p. 23a.

51. Gu, *Chouban Pingxiang tielu gong du*, chap. 4, pp. 23a–24a.

52. There are many studies that make reference to the fragmentation of landholdings. See, for example, Mark Elvin, *Pattern of the Chinese Past* (Stanford: Stanford University Press, 1973), 253; Linda Gale Arrigo, "Landownership Concentration in China: The Buck Survey Revisited," *Modern China* 12, 3 (1986): 277; and Chao Kang, *Man and Land in Chinese History: An Economic Analysis* (Stanford: Stanford University Press, 1986), 118–21. For more on surface and subsurface rights, see Philip C. C. Huang, *The Peasant Family and Rural Development in the Yangzi Delta, 1350–1988* (Stanford: Stanford University Press, 1990), 42, and Kathryn Bernhardt, *Rents, Taxes, and Peasant Resistance: The Lower Yangzi Region, 1840–1950* (Stanford: Stanford University Press, 1992), 21–27. For cases of deposits and surface rights near Pingxiang County, see Peter Perdue, *Exhausting the Earth: State and Peasant in Hunan, 1500–1850* (Cambridge: Harvard University Press, 1987), 154–57.

53. For more on the semi-formal political order of villages and townships, see Hsiao Kung-chuan, *Rural China: Imperial Control in the Nineteenth Century* (Seattle: University of Washington Press, 1960), 43–143; and Prasenjit Duara, *Culture, Power, and the State: Rural North China, 1900–1942* (Stanford: Stanford University Press, 1988), 42–57.

54. Huang, *Peasant Family and Rural Development*, 156–57. Huang notes that subsoil rights were regularly bought and sold by land speculators in the Yangzi delta, which might suggest that the patronage relationships were already tenuous. Freedman, in his early study of Guangdong Province, on the other hand, noted that corporate lineage properties tended to be stable influences over the community. Osborne suggests the same was true for eastern Jiangxi Province. See Maurice Freedman, *Lineage Organization in Southeastern China* (New York: Humanities Press, 1965), 11–12; and Anne Rankin Osborne, "Barren Mountains, Raging Rivers: The Ecological and Social Effects of Changing Land Use in the Yangzi Periphery in Late Imperial China," PhD dissertation (Columbia University, 1989), 113–21.

55. Terry F. Kleeman, "Land Contracts and Related Documents," in *Chugoku no shukyo shiso to kagaku: Makio Ryokai Hakushi shoju kinen ronshu*, ed. Makio Ryokai (Tokyo: Kokusho Kankohkai, 1984); Valerie Hansen, *Negotiating Daily Life in Traditional China: How Ordinary People Used Contracts, 600–1400* (New Haven: Yale University Press, 1995), 24–33.

56. Huang, *Peasant Family and Rural Development*, 106; Jamieson, *Chinese Family and Commercial Law*, 88.

57. Myron L. Cohen, "Writs of Passage in Late Imperial China: The Documentation of Practical Understandings in Minong, Taiwan," in *Contract and Property in Early Modern China*, ed. Madeline Zelin, Jonathan Ocko, and Robert Gardella (Stanford: Stanford University Press, 2004), 40–43; Philip C. C. Huang, *Code, Custom, and Legal Practice in China: The Qing and Republic Compared* (Stanford: Stanford University Press, 2001), 71–98; Jamieson, *Chinese Family and Commercial Law*, 89; Richard John Lufrano, *Honorable Merchants: Commerce and Self-Cultivation in Late Imperial China* (Honolulu: University of Hawaii Press, 1997), 136; Yang Guozhen, *Ming Qing tu di qi yue wenshu yanjiu* (Beijing: Renmin chubanshe, 1988).

58. Contract #101.

59. Contract #10-A.

60. Contract #2-7-131. For more on the claim of hardship as a legitimizing narrative found in the contract, see Cohen, "Writs of Passage in Late Imperial China," 48.

61. Gu, *Chouban Pingxiang tielu gong du*, chap. 2, pp. 26a–26b, italics added.

62. Cohen, "Writs of Passage in Late Imperial China," 46; Thomas M. Buoye, "Litigation, Legitimacy, and Lethal Violence: Why Country Courts Failed to Prevent Violent Disputes over Property in Eighteenth-Century China," in *Contract and Property in Early Modern China*, ed. Madeline Zelin, Jonathan Ocko, and Robert Gardella (Stanford: Stanford University Press, 2004), 99–100; Peter Hoang, "Practical Treatise on Legal Ownership," *Journal of the China Branch of the Royal Asiatic Society of Great Britain and Ireland* 23 (1889): 132–33; Madeline Zelin, *The Magistrate's Tael: Rationalizing Fiscal Reform in Eighteenth-Century Ch'ing China* (Berkeley: University of California Press, 1984), 250–53.

63. Liu, *Zhaoping zhilue*, 1550.

64. Gu, *Chouban Pingxiang tielu gong du*, chap. 4, pp. 3a–3b. The *bao* leader was often a neighborhood or lineage head who oversaw the local militia and served in various semi-official capacities. For more on this, see Hsiao, *Rural China*, 43–83.

65. Elvin, *Pattern of the Chinese Past*, 253.

66. Perdue, *Exhausting the Earth*, 149–50; William C. Jones, *The Great Qing Code* (Oxford: Clarendon Press, 1994), 118–20.

67. E. T. Williams, *Recent Chinese Legislation Relating to Commercial, Railway, and Mining Enterprises with Regulations for Registration of Trade Marks, and for the Registration of Companies* (Shanghai: Shanghai Mercury, 1905), 48, 53–54.

68. One source stated that County Magistrate Gu began with an initial purchase of 333 *mu* of land. See Zhang et al., *Jiangxi sheng zhi*, 10. However, this figure seems much too low for the total length of the rails and may refer to the track from Anyuan to the county seat. Putting together rough numbers for analysis, I determined that the railroad line from Anyuan to the county's western border was about 15 miles long, and must have been at least 20 feet wide. Therefore, Gu would have to purchase nearly 2,000 *mu* of property for the railroad. (Specifically, the total required purchase was about 327 acres or 1962 *mu*.)

69. Gu, *Chouban Pingxiang tielu gong du*, chap. 1, pp. 2a–2b. Arrigo reminds us that the most important aspect of property is indeed location. Specifically, the most lucrative lands were those close to markets and along river routes where cheap transportation costs expanded the reach of the local product. See Arrigo, "Landownership Concentration," 281.

70. The local gazetteer writes that 1 *ba* is equal to 30 *mu*, which will further cloud Gu's calculations. See Liu, *Zhaoping zhilue*, 2357.

71. Gu, *Chouban Pingxiang tielu gong du*, chap. 1, pp. 2a–2b. A picul is 133.33 pounds or 60.477 kilograms. See Frederick Wakeman Jr., *The Great Enterprise: The Manchu Reconstruction of Imperial Order in Seventeenth-Century China* (Berkeley: University of California Press, 1985), xiii.

72. Gu, *Chouban Pingxiang tielu gong du*, chap. 2, p. 26a.

73. Gu, *Chouban Pingxiang tielu gong du*, chap. 1, pp. 2a–2b. Using more universal calculations—based on the many loose conversion tables of Chinese measurements into Western measurements—Gu's findings were that every hectare, about 15.6 *mu*, produced 10,189 kilos of grain (alternately, every acre of land, about 6 *mu*, likely produced over 8,600 pounds).

74. For this conversion rate, see Jerome Ch'en, *Highlanders of Central China: A History, 1895–1937* (Armonk: M. E. Sharpe, 1992), xix.

75. Ch'en, *Highlanders of Central China*, 68. For another set of calculations of land values in taels, see Chao, *Man and Land in Chinese History*, 130.

76. However, when I compare these output numbers with the Wen genealogy—an undated collection that was probably completed in the 1880s—I conclude that Gu's numbers were almost certainly excessive by a factor of two or three. One Wen lineage ancestral property was listed as 515 *ba* in size and took in an annual rent of 50 piculs or *shi*. This calculates as 1.7 hectares of land providing rents of 3023.85 kilos. Assuming rents represent about half the total summer crop output—a fair assumption for late Qing China—we would calculate that the Wen property produced over 6,000 kilos per hectare at fall harvest. Another property was listed as 500 *ba* or 1.6 hectares but took in 40 piculs, or just over 2,419 kilos for a calculated 5,000 kilos per hectare for the first and most significant crop. Even though the Wen lineage numbers suggest less than half the output Gu determined, they are even higher than calculations recently done for this area by a scholar working with John Lossing Buck's 1930s data. In this study the author calculated that on average 1 hectare probably produced only 2,731.3 kilos of grain per hectare. See Arrigo, "Landownership Concentration in China," 320.

77. A cash (*qian*) is 1/5 of a *liang* or a tael in the old Chinese money system, and Feuerwerker writes that 1 Mexican silver dollar was .72 taels. See Albert Feuerwerker, "Economic Trends in the Late Ch'ing Empire, 1870–1911," in *The Cambridge History of China, Volume 11, Part 2: Late Ch'ing, 1800–1911*, ed. Denis Twitchett and John King Fairbank (Cambridge: Cambridge University Press, 1980), 29. Given that the land necessary is about 2,000 *mu* (see note 68 in this chapter) and that each *mu* was to be 60,000 cash, I calculate the total cost to be about 24 million taels or 17,280,000 million Mexican silver dollars. Again,

calculations such as these should be viewed with caution and are meant to suggest the price, not to calculate it with certainty.

78. Gu Jiaxiang, *Lutang wen ji* (Taibei: Wenhai chubanshe, 1972), 285–89.

79. Ch'u Tung-tsu, *Local Government in China under the Ch'ing* (Stanford: Stanford University Press, 1962), 132.

80. Gu, *Chouban Pingxiang tielu gong du*, chap. 1, p. 2a.

81. Gu, *Chouban Pingxiang tielu gong du*, chap. 1, pp. 1a–5b; chap. 2, pp. 26a–28a.

82. For further discussion of collaborators under the stress of imperialist powers, see Allen Isaacman and Barbara Isaacman, "Resistance and Collaboration in Southern and Central Africa, ca. 1850–1920," *International Journal of African Historical Studies* 10, 1 (1977): 31–62.

83. Gu, *Chouban Pingxiang tielu gong du*, chap. 2, pp. 1a–5a.

84. Yan Bing, "Hei an de niandai: bu qu de douzheng," in *Lishi zhishi congshu: Anyuan lukung gongren julebu shihua* (Nanchang: Jiangxi renmin chubanshe, 1983), 2.

85. Contract #10-A

86. Ch'en, *Highlanders of Central China*, 68. This contract does not state that the silver needed to be pure Mexican. Rather, the contract stipulated that the purity was to be agreed upon by the buyer and seller. Since the conversion rate of Chinese taels to foreign currency was about 72 to 100, then even if we assume that the payment was in pure Mexican silver, the payment would be a paltry 10.56 *yuan* per *mu*. For this conversion rate, see Thomas G. Rawski, *Economic Growth in Prewar China* (Berkeley: University of California Press, 1989), xv.

87. Contract #123.

88. Rudolf Hommel, *China at Work: An Illustrated Record of the Primitive Industries of China's Masses, Whose Life Is Toil, and Thus an Account of Chinese Civilization* (New York: John Day Company, 1937), 4.

89. Parsons, *American Engineer in China*, 264–65.

90. Contract #289. However, if we examine this price against Gu's scheme, it would be necessary, given this compensation rate, for the property to be worth no more than the equivalent of 2 *mu* of irrigated farmland if we assume there was no extra compensation for the removal of the ancestors in the ground. Given the size of a normal gravesite in China, this is certainly possible, but it can be no more than pure speculation.

Chapter 5. Mechanization of the Coalmines: Tearing Down and Building Up

1. For descriptions of the so-called first and second industrial revolutions, see, for example, David S. Landes, *The Unbound Prometheus: Technological Change and Industrial Development in Western Europe from 1750 to the Present* (Cambridge: Cambridge University Press, 1969), 231–358; Peter N. Stearns, *The*

Industrial Revolution in World History (Boulder: Westview Press, 1998), 21–32, 111–16.

2. See, for example, Christoph Campregher, "Shifting Perspectives on Development: An Actor-Network Study of a Dam in Costa Rica," *Anthropological Quarterly* 83, 4 (2010): 783–804.

3. In his discussion of civil engineering, Reuss explains that "engineers often spend more time negotiating than building. . . . Few engineers look forward to this process, which inevitably delays construction and creates new uncertainties, but laws and policy prescribe it and it is ignored at agency peril." He further states that among the broad outlines of this negotiation are essentially the conflicts between Scott's "High Modernism," and the vision of a technological order based on Western science, and the *mētis*, or local methodology employed for a similar outcome. Martin Reuss, "Seeing Like an Engineer: Water Projects and the Mediation of the Incommensurable," *Technology and Culture* 49, 3 (2008): 531–46; and James Scott, *Seeing Like a State: How Certain Schemes to Improve the Human Condition Have Failed* (New Haven: Yale University Press, 1998).

4. The literature on this topic is primarily based on political scientists and sociologists whose presentist analysis should be used here with some trepidation. However, it does appear that the post–World War II studies provide us with some acceptable points of comparison. See Francis Fukuyama and Sanjay Marwah, "Dimensions of Development," *Journal of Democracy* 11, 4 (2000): 80–94; Francis Fukuyama, "The Imperative of State-Building," *Journal of Democracy* 15, 2 (2004): 17–31; Gary Gereffi, "Development Strategies and the Global Factory," *Annals of the American Academy of Political and Social Science* 505, The Pacific Region: Challenges to Policy and Theory (1989): 92–104; Stephan Haggard and Steven B. Webb, "What Do We Know about the Political Economy of Economic Policy Reform?" *The World Bank Research Observer* 8, 2 (1993): 143–68.

5. Anthropological studies of development and modernization, for example, point to the sociological tensions inherent in technological change. In her studies of Chinese families and production, Gates describes the lineage as a "patri-corporation" that uses kinship to organize labor and resources toward community benefit and expansion of Confucian and gentry power. However, when factories were established in modern Taiwan, she noticed that peasants who sought work in manufacturing were coerced to leave the safety net of the lineage. Factories, in turn, primarily hired young men and women who did not have children and therefore did not require an income that included reproduction costs. Instead, those costs were placed at the feet of the lineages and villages, even as those institutions were losing some of their most productive laborers. Thus, as peasant families engaged in greater self-exploitation to maintain their subsistence, workers turned to contract labor bosses and secret societies in the hopes of hanging on to the paternal bonds that provided them with greater assurance of survival back home. See Hill Gates, "Dependency and the Part-Time Proletariat in Taiwan," *Modern China* 5, 3 (1979): 381–407.

6. For this argument, see, for instance, Frederic C. Deyo, "Labor and Development Policy in East Asia," *Annals of the American Academy of Political*

and Social Science 505 The Pacific Region: Challenges to Policy and Theory (1989): 152–61.

7. One Western observer wrote, "Too much praise cannot be accorded to the foreign engineers, who are all German, headed by Gustavus Leinung, for what they have accomplished in the building of this plant and the difficulties they have overcome." See K. P. Swensen, "The Pinghsiang Colliery," *Mining and Scientific Press* (October 29, 1910): 564.

8. Charlton M. Lewis, *Prologue to the Chinese Revolution: The Transformation of Ideas and Institutions in Hunan Province, 1891–1907* (Cambridge: East Asian Research Center, Harvard University Press, 1976), 127; Liu Hongpi, ed., *Zhaoping zhilue* (Taiwan: Chengwen chubanshe, 1975), 893.

9. Liu, *Zhaoping zhilue*, 81.

10. For discussions of various spatial models of premodern Chinese marketing, see Kwan Man Bun, "Mapping the Hinterland: Treaty Ports and Regional Analysis in Modern China," in *Remapping China: Fissures in Historical Terrain*, ed. Gail Hershatter, Emily Honig, Jonathan Lippman, and Randall Stross (Stanford: Stanford University Press, 1996); Carol Smith, ed., *Regional Analysis: Volume I Economic Systems* (New York: Academic Press, 1976), and Carol Smith, *Regional Analysis: Volume II Social Systems* (New York: Academic Press, 1976); and G. William Skinner, "Marketing and Social Structure," Parts I and II, *Journal of Asian Studies* 24, 1 and 2 (1965): 3–43, 195–228.

11. Gu, *Zhongguo shi da kuan changdiao ji* (Shanghai: Shanghai Commercial Press, 1916), chap. 12, p. 3; Alfred C. Reed, "Coal Mining in China," *The Scientific Monthly* 5 (1917): 38.

12. Liu, *Zhaoping zhilue*, 894, 896–99; Changsha shi geming jinian de ban gong tai and Anyuan lukuang gongren yundong jinian guang, eds., *Anyuan lukuang gongren yundong shiliao* (Changsha: Hunan gongren chubanshe, 1980), 512–13 (hereafter referred to as *AYLK*).

13. A *chi* is .3581 meters and is therefore slightly more than 1 U.S. foot in length.

14. Gu, *Zhongguo shi da kuan chang*, chap. 12, p. 3. As was stated in chapter 2, the Zijia Mountain area was initially owned by the Wen lineages, while the Tianzi Mountain was originally owned by the Peng lineage. Though I showed in chapter 4 that the Wens sold their property to the Mine Bureau, it is apparent that the Pengs were forced to sell their mines as well.

15. Chen Xulu and Gu Tinglong, *Hanyeping gongsi*, Vol. 2 (Shanghai: Shanghai renmin chubanshe, 1986), 62.

16. Chen and Gu, *Hanyeping gongsi*, Vol. 2, 61–64.

17. Liu, *Zhaoping zhilue*, 896–99; Tim Wright, *Coal Mining in China's Economy and Society, 1895–1937* (Cambridge: Cambridge University Press, 1984), 38.

18. *AYLK*, 510–17; Gu, *Zhongguo shi da kuan chang*, chap. 3, pp. 21–23; "The Pingsiang Colliery: A Story of Early Mining Difficulties in China," *The Far Eastern Review* 12, 10 (1916): 376; Swensen, "Pinghsiang Colliery," 564. It should be noted that these trains are still running in the mines even as the mines are no longer working.

19. Yan Bing, “Hei an de nian dai: bu qu de douzheng,” in *Lishi zhishi congshu: Anyuan lukuang gongren julebu shihua* (Nanchang: Jiangxi renming chubanshe, 1983), 3; Chen and Gu, *Hanyeping gongsi*, Vol. 2, 123–28.

20. Chen and Gu, *Hanyeping gongsi*, Vol. 2, 78–79.

21. Chen and Gu, *Hanyeping gongsi*, Vol. 2, 123–28.

22. I. C. F. Stratham, *Coal Mining Practice*, Vol. 4 (London: Caxton Publishing Company, 1958), 344-66. While Leinung initially assumed that they could purify 500 tons in ten hours, the two scrubbers used in the coalmine by the middle of the second decade of the twentieth century produced as much as 3,000 tons per day. See Zhonggong Pingxiang meikuang weiyuan hui xian zhuan bubian, eds., *Hongse de Anyuan* (Nanchang: Jiangxi renmin chubanshe, 1959), 27–28 (hereafter referred to as *HDA*); Gu, *Zhongguo shi da kuang chan*, chap. 3, pp. 23–25; Liu, *Zhaoping zhilue*, 895; “Pingsiang,” 377.

23. Chen and Gu, *Hanyeping gongsi*, Vol. 2, 80–81.

24. Chen and Gu, *Hanyeping gongsi*, Vol. 2, 84–86.

25. Reed, “Coal Mining in China,” 45–46.

26. Chen and Gu, *Hanyeping gongsi*, Vol. 2, 123–28.

27. “Pingsiang,” 375–77; Swensen, “Pinghsiang Colliery,” 564–65.

28. *AYLK*, 510, 513.

29. *HDA*, 14–15.

30. T. F. Hou, *General Statement on the Mining Industry*, Vol. 5 (Beijing: Geological Survey of China, 1935), 487–88.

31. Luo Xiao, *Pingxiang shi difang meitan gongye zhi* (Nanchang: Jiangxi renmin chubanshe, 1992), 49; Gu, *Zhongguo shi da kuang chan*, chap. 3, pp. 8–9; Lansing W. Hoyt, “Blast Furnaces and Steel Mills in China: A Comprehensive Survey of China's Steel Industry,” *The Far Eastern Review* 19, 5 (1923): 315.

32. See Joseph W. Esherick, *Reform and Revolution in China: The 1911 Revolution in Hunan and Hubei* (Berkeley: University of California Press, 1976), 58–65, for brief references to coalmining in Anle Township.

33. Thomas G. Rawski, *Economic Growth in Prewar China* (Berkeley: University of California Press, 1989), 181–238.

34. V. K. Lee, “China's Iron and Steel Industry,” *The Chinese Students Monthly* 9, 4 (1914): 287; Liu, *Zhaoping zhilue*, 368–69; Fu Xiongxiang, *Liling xiangtu zhi* (Taiwan: Chengwen chubanshe, 1926), 31; Percy Horace Kent, *Railway Enterprise: An Account of Its Origin and Development* (London: Edward Arnold, 1907), 112–15, 121. For a map and brief description of the branch line route, see Zhang Yi, Zhu Yuanqing, and Fan Yinfei, eds. *Jiangxi sheng zhi* (Beijing: Zhonggong zhong yang dang xiao chubanshe, 1993), 9–10.

35. Lewis, *Prologue of the Chinese Revolution*, 124.

36. Percy M. Roxby, “Wu-Han: The Heart of China,” *Scottish Geographical Magazine* 32 (1916): 275–76.

37. Geo Bronson Rea, “Railway Loan Agreements and Their Relations to the Open Door: A Plea for Fair Play to China,” *The Far Eastern Review* 6, 6 (1909): 225–26.

38. Walworth Tyng, "The Miners' Church at Peaceful Spring: Among Collieries and Coke Ovens at Anyuen—A Vivid Picture of Our Work in a Little Known Par of the District of Hankow," *The Spirit of the Missions* 90 (1925): 477. One source indicates that in 1905 the trains only traveled twice per day. See Liu Minghan and Ma Jingyuan, eds., *Hanyeping gongsi zhi* (Wuhan: Huazhong li gong daxue chubanshe, 1990), 70.

39. After the entire railroad was completed, one Western observer calculated that the coke that cost $2.50 gold per ton in Anyuan brought in a price of $14 gold per ton in Hanyang. Even more, coke required in Daye—which increased greatly in the 1910s—would have been even higher due to its greater distance from the coalmine plant gate. For these calculations, see Hoyt, "Blast Furnaces and Steel Mills in China," 317.

40. Fu, *Liling xiangtu zhi*, 66; Gu, *Zhongguo shi da kuang chan*, chap. 3, p. 20; Stephen Carl Averill, "Revolution in the Highlands: The Rise of the Communist Movement in Jiangxi Province," PhD dissertation (Cornell University, 1982), 20–24.

41. Fu, *Liling xiangtu zhi*, 89.

42. Lewis, *Prologue to the Chinese Revolution*, 122.

43. Chang Kuo-t'ao, *The Rise of the Chinese Communist Party, 1921–1927: The Autobiography of Chang Kuo-t'ao* (Lawrence: University of Kansas Press, 1971), 16.

44. Liu, *Zhaoping zhilue*, 41.

45. Nym Wales, *The Chinese Communists: Sketches and Autobiographies of the Old Guard* (Westport: Greenwood, 1972), 90.

46. For more on the impact of railroads on local markets and transportation, see Francis T. Evens, "Roads, Railways, and Canals: Technical Choices in Nineteenth-Century Britain," in *Technology and the West: A Historical Anthology from "Technology and Culture,"* ed. Terry S. Reynolds and Stephen H. Cutcliffe (Chicago: University of Chicago Press, 1997); Smith, *Regional Analysis I* and *Regional Analysis II*; and Kenneth Pomeranz, *Making of a Hinterland: State Society and Economy in Inland North China, 1853–1937* (Berkeley: University of California Press, 1993).

47. Chang, *Rise of the Chinese Communist Party*, 16–17.

48. For brief discussions of the economy in Lukou, see Fu, *Liling xiangtu zhi*, 32; Lewis, *Prologue of the Chinese Revolution*, 144

49. "Pingsiang," 378; Jiangxi sheng Pingxiang shi zong gong hui and Anyuan lukuang gongren yundong ji hui guang, eds., *Anyuan lukuang gongren dabao shengli liushi zhou nian ji hui hua ce* (Jiangxi: N.p., 1982), 23 (hereafter referred to as *AYBGSL*).

50. Angus McDonald, *Urban Origins of the Rural Revolution: Elites and the Masses in Hunan Province, China, 1911–1927* (Berkeley: University of California Press, 1978), 166; *HDA*, 14–15.

51. L. C. Arlington, *Through the Dragon's Eyes: Fifty Year's Experiences of the Foreigner in the Chinese Government Services* (New York: Richard R. Smith, 1931), 196.

52. Boris Torgasheff, *Mining Labor in China* (Shanghai: Bureau of Industrial and Commercial Information, Ministry of Industry, Commerce and Labor, and National Government of the Republic of China, 1930), 53–60.

53. "Pingsiang," 378–79.

54. Lynda Shaffer, *Mao and the Workers: The Hunan Labor Movement, 1920–1923* (Armonk: M. E. Sharpe, 1982), 81; Gu, *Zhongguo shi da kuang chan*, chap. 3, 54; *AYLK*, 30.

55. Gu, *Zhongguo shi da kuang chan*, chap. 3, p. 54.

56. Liu Zongdao, "Lishi pengbei wandai yongcun: fang Anyuan sanji," *Dang de shenghuo congkan* 3 (1980): 37–38.

57. Shaffer, *Mao and the Workers*, 81, 84.

58. See Torgasheff, *Mining Labor in China*, 57. Torgasheff's study finds that feeding workers was nearly always the specific responsibility of the contract labor boss.

59. More on this will be discussed in the next chapter.

60. Swensen explains, "The greatest difficulty is found in the maintenance of haulage and passageways, because of the tremendous pressure. This makes timbering an expensive item, for the timbers must be constantly replaced. No attempt is made to relieve the pressure on the main timbers by providing auxiliary sets as is sometimes done for swelling ground in America. The roof is allowed to fall until it becomes low enough to interfere with the workmen, when it is cut down and new sets are put in." Swensen, "Pinghsiang Colliery," 566.

61. Yan, "Hei an de nian dai," 4; Albert Feuerwerker, "China's Nineteenth-Century Industrialization: The Case of the Hanyeping Coal and Iron Company, Limited," in *The Economic Development to China and Japan: Studies in Economic History and Political Economy*, ed. C. D. Cowan (New York: Frederick Praeger, 1964), 94.

62. Shaffer shows that in other Chinese mines at this time, while several hundred were reportedly killed, the number hurt were at times put in the thousands. See Shaffer, *Mao and the Workers*, 75, 78–79; Gu, *Zhongguo shi da kuan chan*, chap. 3, pp. 11–12.

63. Shaffer, *Mao and the Workers*, 79; Jean Chesneaux, *Chinese Labor Movement, 1919–1927*, trans. H. M. Wright (Stanford: Stanford University Press, 1968), 79; Wright, *Coal Mining in China's Economy and Society*, 175. For more on hookworms, see Peter J. Horetz, "China's Hookworms," *The China Quarterly* 172 (2002): 1030–32; W. A. Sawyer, "Hookworm Disease as Related to Industry in Australia," *The American Journal of Tropical Medicine* 327, 3 (1923): 159–69; "Hookworm Disease in China," *British Medical Journal* 1, 3460 (1927): 807.

64. "Pingsiang," 378.

65. Lewis, *Prologue to the Chinese Revolution*, 127.

66. In her recent study, Elizabeth Perry writes that this hospital was an impressive facility. She explains that it was "fully equipped" and included as many as twenty doctors, both Chinese and foreign, to attend to the miners' needs. However, the myriad troubles faced by the clinic doctors may have been more

than they could handle. Wright, for example, points out that the medical staffs of the mines in China were hired specifically for mending laborers who were hurt in accidents. The high numbers of cases of illness due to unsanitary labor conditions described earlier were therefore probably not properly treated. See Elizabeth J. Perry, *Anyuan: Mining China's Revolutionary Tradition* (Berkeley: University of California Press, 2012), 19; Wright, *Coal Mining in the China's Economy and Society*, 175.

67. Michael T. Taussig, for example, points out that in colonial-era Latin America the European doctors were seen to rely as much on their Christian faith as their scientific skills and medical procedures. He suggests that curing the patient often involved prayer and rosary beads as much as pills and technology. See Michael T. Taussig, *Devil and Commodity Fetishism in South America* (Chapel Hill: University of North Carolina Press, 1980), 41–46. See also Ellsworth C. Carlson, *The Kaiping Mines, 1877–1912* (Cambridge: Harvard University Press, 1971), 47; and *AYLK*, 510.

68. Ping Wen, "Zhang Zanchen jian jie," in *Pingxiang meitan fazhan shilue*, ed. Jiangxi sheng zheng xi wenshi ziliao yanjiu weiyuan and Pingxiang shi zheng wei wenshi ziliao yanjiu weiyuan hui he bian (Hong Kong: N.p., 1987), 194–95.

69. Swensen, "Pinghsiang Colliery," 565–66.

70. Liu, *Zhaoping zhilue*, 1084; *AYLK*, 514–16.

71. Perry, *Anyuan*, 38–39; Tyng, "Miner's Church at Peaceful Spring."

72. Chen Xulu and Gu Tinglong, *Hanyeping gongsi*, Vol. 1 (Shanghai: Shanghai renmin chubanshe, 1984), 230–32, 235.

73. Oddly enough, there are no documents that call the building the Sheng Xuanhuai Memorial Hall. However, a plaque at the entrance indicates that this is the official name of the building. It is unclear when this name was bestowed upon the building.

74. Chang Sen-dou, "Some Aspects of the Urban Geography of the Chinese Hsien Capital," *Annals of the Association of American Geographers* 51, 1 (1951): 37; Chang Sen-dou, "The Morphology of Walled Capitals," in *The City in Late Imperial China*, ed. G. William Skinner (Stanford: Stanford University Press, 1977), 99; and Nancy Shatzman Steinhardt, *Chinese Imperial City Planning* (Honolulu: University of Hawaii Press, 1990), ix–x. Each of these sources indicate that the concept of a perfect county seat with exact planning was more a bureaucratic desire than reality.

75. "Pingsiang," 378; and Anthony D. King, *The Bungalow: The Production of a Global Culture* (London: Routledge and Kegan Paul, 1984).

76. G. Anthony Atkinson, "British Architects in the Tropics," *Architectural Association Journal* 69 (1953): 7–21. The low ceilings on the way up the stairs reminded the author that the German engineer Gustav Leinung was not a very tall man.

77. Frederick Cooper, "Urban Space, Industrial Time, and Wage Labor in Africa," in *Struggle for the City: Migrant Labor, Capital, and the State in Urban Africa*, ed. Frederick Cooper (Beverly Hills: Sage, 1983), 31.

78. Li Jui, *Early Revolutionary Activities of Comrade Mao Tse-tung* (White Plains: M. E. Sharpe, 1997), 201.

79. *AYLK*, 30; and "Pingsiang," 377.

80. Fei-ling Davis, *Primitive Revolutionaries of China: A Study of Secret Societies of the Late Nineteenth Century* (Honolulu: University of Hawaii Press, 1971), 86.

81. *AYLK*, 514–15. Perry writes that the park was Western-styled and suitable. See Perry, *Anyuan*, 20.

82. Torgasheff explains that contract labor bosses agreed to provide free coffins to all workers killed while working for them. See Torgasheff, *Mining Labor in China*, 56; and *HDA*, 35.

83. Chen and Gu, *Hanyeping gongsi*, Vol. 1, pp. 572–73.

84. Chen and Gu, *Hanyeping gongsi*, Vol. 1, pp. 446–447.

85. Chen and Gu, *Hanyeping gongsi* Vol. 2, pp. 501–02.

86. Shaffer, *Mao and the Workers*, 81–82.

87. "Pingsiang," 380; Wales, *Chinese Communists*, 85; and *HDA*, 16–28.

88. "Pingsiang," 377.

89. *HDA*, 16–28.

90. In other cases of industrialization this solution was resolved, though uneasily, using foreigners and ethnic minorities. As industry grew in the United States, for example, workers were brought in from other countries where semi-skilled labor could be found. In some cases communities of people from ethnic and racial minorities were used to do difficult and dangerous labor. See, for example, Joe William Trotter, *Coal, Class, and Color: Blacks in Southern West Virginia, 1915–32* (Urbana: University of Illinois Press, 1990).

91. "Pingsiang," 377.

92. Sun E-tu Zen, "Mining Labor in the Ch'ing Period," in *Approaches to Modern Chinese History*, ed. Albert Feuerwerker (Berkeley: University of California Press, 1967), 54.

93. Ferdinand Paul Wilhelm Freiherr von Richthofen, *Baron Richthofen's Letters, 1870–1872* (Shanghai: North China Herald, n.d.), 4, 10.

94. Chen and Gu, *Hanyeping gongsi*, Vol. 1, pp. 603–05.

95. Perry, *Anyuan*, 21.

96. Esherick, *Reform and Revolution in China*, 58–62.

97. "Pingsiang," 378; and Shaffer, *Mao and the Workers*, 80.

98. Shaffer, *Mao and the Workers*, 82–83.

99. Wales, *Chinese Communists*, 83.

100. For a similar depiction of this type of recruitment, see Gail Hershatter, *Workers of Tianjin, 1900–1949* (Stanford: Stanford University Press, 1986), 51.

101. *HDA*, 15–22.

102. Richard Perez-Pena, "Gen. Yang Dezhi, 83, Dies: Was Chinese Army Chief," *New York Times* (28 October 1994): A-15; Union Research Institute, *Who's Who in Communist China* (Hong Kong: Union Research Institute, 1969), 756.

103. Leonard G. Ting, "Coal Industry of China," *Nankai Social and Economic Quarterly* 10, 2 (1937): 249.

104. Swensen, "Pinghsiang Colliery," 567.

105. Torgasheff, *Mining Labor in China*, 7, 54–56.

106. In any case, because Chinese currency during the late Qing Dynasty and the Republican period fluctuated greatly, it is difficult to determine the actual wage rates of the workers. For instance, one miner explained that he made no more than 7 *yuan* (about 5 taels) a month, which is similar to another source that placed daily wages at no more than .3 *yuan* a day. Another miner explained that he made between 120 and 1,000 cash per day over his tenure in the mines. A Western observer stated that miners were paid "less than 13 cents gold per day." Since these currencies often fluctuated, it would be hard to provide an exact conversion for this price. See Shaffer, *Mao and the Workers*, 76–78; Wales, *The Chinese Communists*, 83–84; *AYLK*, 28–29; and Swensen, "Pinghsiang Colliery," 566.

107. Chen and Gu, *Hanyeping gongsi*, Vol. 2, pp. 86–89.

108. Arlington, *Through the Dragon's Eyes*, 196.

109. *AYLK*, 516.

110. Chen and Gu, *Hanyeping gongsi*, Vol. 2, pp. 501–02.

111. Wales, *Chinese Communists*, 96.

112. Wright, *Coal Mining in China's Economy and Society*, 178. For a similar discussion of this type of labor in South Africa, see Charles van Onselen, *New Babylon, New Nineveh: Everyday Life on the Witwatersrand, 1886–1914* (Johannesburg: Jonathan Ball, 1982).

113. *AYLK*, 514; Chang, *Rise of the Chinese Communist Party*, 16.

114. "Pingsiang," 377.

115. Arlington, *Through the Dragon's Eyes*, 196.

116. *HDA*, 15–22.

117. Kimio Nishizawa, "Mines in the Yangtze Valley, China," *Journal of the Royal Society of Arts* 58 (1910): 944.

118. Gail Hershatter shows that in the much more industrialized and metropolitan city of Shanghai, for example, of those Chinese who migrated to the Chinese sections of the city, the ratio of men to women was about 135 to 100 during the 1930s, a ratio almost certainly less skewed than that found in the mining town of Anyuan. Yet she uses these numbers to argue that the lack of prospective female companionship sparked demand for prostitution in the city. Gail Hershatter, "Hierarchy of Shanghai Prostitution, 1870–1949," *Modern China* 15, 4 (1989): 465–75.

119. *AYLK*, 30.

120. Esherick, *Reform and Revolution*, 59–60.

121. *HDA*, 22.

122. Coked coal output did not rise as high as raw coal as it was listed as 29,000 tons in 1898, 215,765 in 1910, and only 166,062 in 1911. Liu and Ma, *Hanyeping gongsi zhi*, 75.

123. Chen and Gu, *Hanyeping gongsi*, Vol. 2, 558–59. For more on Sheng Xuanhaui's loans with Japanese banking firms, see Zhong yang yanjiuyuan and Jindai shi yanjiusuo, *Kuang wu dang*, Vol. 4: Anhui, Jiangxi, Hubei, and Hunan Provinces (Taibei: Zhong yang yanjiuyuan jindai shi yanjiusuo, 1960), 2285–2302 (hereafter referred to as *KWD*).

124. Chen and Gu, *Hanyeping gongsi*, Vol. 2, pp. 642–43, 674–76.

125. Feuerwerker, "China's Nineteenth-Century Industrialization," 107.

126. F. R. Tegengren, *Iron Ores and Iron Industry of China*, Part II (Beijing: Geological Survey of China, Ministry of Agriculture and Commerce, 1923–1924), 368; Mauris B. Jansen, "Yawata, Hanyehping, and the Twenty-One Demands," *Pacific Historical Review* 23, 1 (1954): 35; William F. Collins, *Mineral Enterprise in China* (New York: Macmillan, 1918), 85; and Frederick R. Dickinson, *War and National Reinvention: Japan and the Great War, 1914–1919* (Cambridge: Harvard University Press, 1999), 89.

127. Collins, *Mineral Enterprise in China*, 83.

128. Tegengren, *Iron Ores and Iron Industry of China*, 379.

129. Jansen, "Yawata, Hanyehping, and the Twenty-One Demands," 42; and Dickinson, *War and National Reinvention*, 89.

130. "Notes on Coal Mines in China," *The Far Eastern Review* 37 (1931): 203; Feuerwerker, "China's Nineteenth-Century Industrialization," 90. Tegengren, *Iron Ore and Iron Industry of China*, 370, places the losses to the company from the brief shutdown due to the 1911 Revolution alone at 3.7 million taels.

131. Norman D. Hanwell, "Tayeh Iron Fields Put under Direct Japanese Control," *Far Eastern Survey* 8, 4 (1939): 47.

132. K. L. Hsueh, "Iron and Steel Industry of China," *The Chinese Economic Journal* 2, 1 (1928): 14. Read suggested that a more intensive use of mining technology in the future might lead to greater iron ore output than even Leinung imagined. Specifically, he mentioned that the iron ore deposits that Leinung and others were working with were only the surface deposits and that much more mineral was still to be excavated in lower depths. Thomas T. Read, "The Mineral Production and Resources of China," *Transactions of the American Institute of Mining Engineers* 43 (1912): 33.

133. "Pingsiang," 376.

134. Hoyt, "Blast Furnaces," 315; and "Pingsiang," 376.

135. Hoyt, "Blast Furnaces," 315.

136. While official documents state that Pingxiang County continued to reach tens of millions of tons of coal output annually, the mines in Anyuan had few workers moving shale around. Today, no coking ovens fire and no scrubbers are operational. See Huang Shiguo, ed., *Pingxiang shi zhi* (Pingxiang: Pingxiang shi zhi bianzuan weiyuan huibian gangzhi chubanshe chuban, 1996), 186–190.

137. H. C. Huggins, "Japan's Iron and Steel Industry," *The Far Eastern Review* 19, 3 (1923): 190; and Collins, *Mineral Enterprise in China*, 80.

138. Quoted in Collins, *Mineral Enterprise in China*, 86.

139. Jansen, "Yawata, Hanyeping, and the Twenty-One Demands," 46–48; and Lee, "China's Iron and Steel Industry."

140. "Notes on Coal," 202.

141. H. G. W. Woodhead, *The China Year Book 1926–7* (Chicago: University of Chicago Press, 1927), 111; Tegengren, *Iron Ores and Iron Industry in China*, 370–71, 379–80. See also "The Present Financial Status of the Hanyehping Company, with a List of Outstanding Foreign Loans," *The Far Eastern Review* 23, 7 (1927): 329, which places the total value of Pingxiang Coalmines at only slightly under $11 million.

142. Walter J. Ballard, "As Our Consultants Say," *Los Angeles Times* (9 April 1908): 115.

143. Feuerwerker, "China's Nineteenth-Century Industrialization," 92; and T. F. Hou, *General Statement on the Mining Industry*, Vol. 3 (Beijing: Geological Survey of China, 1935), 130–32. The findings in this publication show that Pingxiang County's cost of coal per ton was about five times as high as the Western-controlled Kailuan mines in northeastern China (see 247–48).

144. "Present Financial Status of the Hanyeping Company," 329.

145. Jansen, "Yawata, Hanyehping, and the Twenty-One Demands," 47–48.

Chapter 6. Social Atomization and Local Resistance: Divergent Desires and Strategies of Elites and Workers

1. E. P. Thompson, *The Making of the English Working Class* (New York: Vintage, 1963).

2. The literature on this for Africa, for example, is extensive. Claude Meillassioux, "From Reproduction to Production: A Marxist Approach to Economic Anthropology," *Economy and Society* 1, 1 (1972): 93–105, argues that workers were separated from their families, thus affecting the means of social reproduction. Other sources, however, indicate that this separation was incomplete and led to rural-urban alliances against European powers. For example, see Charles van Onselen, *Chibaro: African Mine Labour in Southern Rhodesia, 1900–1933* (Johannesburg: Ravan Press, 1980). In China, separation was not as formal, but relationships with the countryside remained fluid and thus affected social class articulation. See, for example, Gail Hershatter, *Workers of Tianjin, 1900–1949* (Stanford: Stanford University Press, 1986).

3. See, for example, Mark Elvin, "Administration of Shanghai, 1905–1919," in *The Chinese City between Two Worlds*, ed. Mark Elvin and G. William Skinner (Stanford: Stanford University Press, 1974); William Rowe, *Hankow: Commerce and Society in a Chinese City, 1796–1889* (Stanford: Stanford University Press, 1984), and William Rowe, *Hankow: Conflict and Community in a Chinese City, 1796–1895* (Stanford: Stanford University Press, 1989).

4. See, for example, Dipesh Chakrabarty, "Conditions for Knowledge of Working-Class Conditions: Employers, Government, and the Jute Workers of

Calcutta, 1890–1940," in *Selected Subaltern Studies*, ed. Ranajit Guha and Gayatri Chakravorty Spivak (New York: Oxford University Press, 1988).

5. Some scholars refer to this new economy as "capitalism" because the forces of production follow Marx's shorthand depiction of that economic stage. I have tried to avoid that problematic term as well as "modernization," another word that may blur more than it reveals. Instead, I show in this chapter that the community changed under the pressure of outside forces and modernization into a newly emerging social and political reality that cannot be easily pigeonholed. For use of these terms, see, Alexander Gerschenkron, *Economic Backwardness in Historical Perspective: A Book of Essays* (Cambridge: Harvard University Press, 1966), 31–51, 68, 177–80.

6. Cooper tells us that the term "Colonialism" has become something of a political football, bandied about by historians and other scholars alike. See Frederick Cooper, *Colonialism in Question: Theory, Knowledge, History* (Berkeley: University of California Press, 2005). Others have sought to problematize the notion of imperialism with phrases like "Formal Empire" versus "Informal Empire." See, for example, Peter Duus, Ramon Hawley Myers, and Mark R. Peattie, eds., *The Japanese Informal Empire in China, 1895–1937* (Princeton: Princeton University Press, 1989). I am not going to enter into this conversation further, but it is apparent that in this case Western power is tangential and based more on economic support and technological know-how rather than military force. At the same time, since the Pingxiang County Coalmines were created in part to stave off real military threats posed by Japan and the Western powers, the role of imperialism in its most brutal form was present.

7. Chakrabarty, "Conditions for Knowledge of Working-Class Conditions."

8. Hua Wen and Luo Xiao, "Pingxiang meitan fazhan shilue," in *Pingxiang meitan fazhang shilue*, ed. Jiangxi sheng zheng xi wenshi ziliao yanjiu weiyuan hui and Pingxiang shi zheng wei wenshi ziliao weiyuan hui he bian (Hong Kong: N.p., 1987), 4.

9. Chen Xulu and Gu Tinglong, eds., *Hanyeping gongsi*, Vol. 1 (Shanghai: Shanghai renmin chubanshe, 1984), 228.

10. Chen and Gu, *Hanyeping gongsi*, Vol. 1, pp. 446–48.

11. Chang Sen-dou calls this area the county seat's Cultural Center as it was dominated by the Confucian Temple and other education-based structures. See Chang Sen-dou, "Urban Geography of the Chinese Hsien Capital," *Annals of the Association of American Geographers* 51, 1 (1951): 40–41.

12. Chen and Gu, *Hanyeping gongsi*, Vol. 1, pp. 230–31.

13. Chen and Gu, *Hanyeping gongsi*, Vol. 1, pp. 230–32.

14. Chen and Gu, *Hanyeping gongsi*, Vol. 1, p. 237.

15. Chen and Gu, *Hanyeping gongsi*, Vol. 1, p. 232.

16. Chen and Gu, *Hanyeping gongsi*, Vol. 1, p. 234.

17. Chen and Gu, *Hanyeping gongsi*, Vol. 1, p. 235.

18. Chen and Gu, *Hanyeping gongsi*, Vol. 1, p. 238.

19. Chen and Gu, *Hanyeping gongsi*, Vol. 1, pp. 228–29.

20. Chen and Gu, *Hanyeping gongsi*, Vol. 1, pp. 234–35.

21. In apparent confirmation of this, no students are listed in the gazetteer as achieving the county-level degree for 1896, a fact that also hindered research in identifying the authors of the posters.

22. Chen and Gu, *Hanyeping gongsi*, Vol. 1, p. 228.

23. Chen and Gu, *Hanyeping gongsi*, Vol. 1, p. 238.

24. Chen and Gu, *Hanyeping gongsi*, Vol. 1, p. 229. In fact, posters attacking the arrival of Gustav Leinung continued to be written years later. Sometime after 1900 a poster was placed on the door to his living quarters that called for his death and the deaths of all who helped him in his mining efforts. Though the poster was apparently unsigned, such posters must have been written by the scholarly elites and indicates further the level of anxiety among the local gentry. See "The Pingsiang Colliery: A Story of Early Mining Difficulties in China," *The Far Eastern Review* 12, 10 (1916): 377.

25. "Pingsiang," 377.

26. Zhongguo shixuehue, ed., *Xinhai geming* (Shanghai: Renmin chubanshe, 1957), 480 (hereafter referred to as *XHGM*).

27. Chang Kuo-t'ao, *The Rise of the Chinese Communist Party, 1921–1927: The Autobiography of Chang Kuo-t'ao* (Lawrence: University of Kansas Press, 1971), 16.

28. This depiction is reminiscent of the integration of demons in tin mines in Bolivia. In his study of these mines, Taussig indicates that the penetration of capitalism into the villages led people to fear the changes imposed upon them. To describe and explain these fears and the concomitant new levels of exploitation, the people used religious iconography that had previously described the farmland in nurturing tones but now depicted the mines as locations of supernatural exploitation and oppression. See Michael T. Taussig, *The Devil and Commodity Fetishism in South America* (Chapel Hill: University of North Carolina Press, 1980), 143–54.

29. Chen and Gu, *Hanyeping gongsi*, Vol. 1, pp. 228–29; Liu Hongpi, ed., *Zhaoping zhilue* (Taiwan: Chengwen chubanshe, 1975), 1481–82.

30. Mechtild Leutner and Andreas Steen, *Deutch-chinesische Beziehungen 1911–1927: vom Kolonialismus zur "Gleichberechtigung": eine Quellensammlung* (Berlin: Akademie Verlag, 2006), 286–91.

31. Chen Xulu and Gu Tinglong, *Hanyeping gongsi*, Vol. 2 (Shanghai: Shanghai renmin chubanshe, 1986), 212–13.

32. Chen and Gu, *Hanyeping gongsi*, Vol. 2, pp. 212–13.

33. Liu, *Zhaoping zhilue*, 1481–82; Fu Xiongxiang, *Liling xiangtu zhi* (Taiwan: Chengwen chubanshe, 1926), 31.

34. Liu, *Zhaoping zhilue*, 1481–82.

35. As has been stated in previous chapters, a picul of grain is calculated at 133.33 pounds.

36. Liu, *Zhaoping zhilue*, 1481–82.

37. Liu, *Zhaoping zhilue*, 896, 1550–51; Quan Hansheng, ed., *Hanyeping gongsi shilue* (Xianggang: Xianggang zhongwen daxue chubanshe, 1972), 211–12.

38. Gu Lang, *Zhongguo shi da kuan chan changdial chaji* (Shanghai: Commercial Press, 1916), chap. 3, p. 46. For more on this, see F. R. Tegengren, *Iron Ores and Iron Industry of China*, Part II (Beijing: Geological Survey of China, Ministry of Agriculture and Commerce, 1923–1924), 270–76.

39. Liu, *Zhaoping zhilue*, 114, 900–03.

40. Joseph W. Esherick, *Reform and Revolution in China: The 1911 Revolution in Hunan and Hubei* (Berkeley: University of California Press, 1976), 27.

41. Gu, *Zhongguo shi da kuan chang*, chap. 3, p. 46.

42. Chen and Gu, *Hanyeping gongsi*, Vol. 2, pp. 327–33.

43. Liu, *Zhaoping zhilue*, 900–03.

44. Liu, *Zhaoping zhilue*, 900–03.

45. Liu, *Zhaoping zhilue*, 903.

46. Gu, *Zhongguo shi da kuan chang*, chap. 3, p. 46.

47. Chen and Gu, *Hanyeping gongsi*, Vol. 1, pp. 194–95.

48. There is a large literature on vandalism and low-level violence as acts of resistance by "the weak." See, for example, James Scott, *Weapons of the Weak: Everyday Forms of Peasant Resistance* (New Haven: Yale University Press, 1985), and James Scott, *Domination and the Arts of Resistance: Hidden Transcripts* (New Haven: Yale University Press, 1990).

49. Chen and Gu, *Hanyeping gongsi*, Vol. 1, pp. 228–29.

50. Samuel Yale Kupper, "Revolution in China: Kiangsi Province, 1905–1913," PhD dissertation (University of Michigan, 1973), 83–92.

51. Lynda Shaffer, *Mao and the Workers: The Hunan Labor Movement, 1920–1923* (Armonk: M. E. Sharpe, 1982), 83.

52. V. K. Lee, "China's Iron and Steel Industry," *The Chinese Students Monthly* 9, 4 (1914): 284, 286.

53. "Pingsiang," 379.

54. "Pingsiang," 378–79. Based on this description, it is interesting to note that the alliance was brought together with the assistance of at least one member of the scholarly elites. Though the previous quotation does not explain whether this person was a member of one of the gentry families or an educated member of the merchant classes, it does provide evidence that some elites were involved in secret society activities or were at least supportive of anti-mining sentiments.

55. "Pingsiang," 379.

56. Zhonggong Pingxiang meikuang weiyuan hui xuan zhuan bubian, eds., *Hongse de Anyuan* (Nanchang: Jiangxi renmin chubanshe, 1959), 27 (hereafter referred to as *HDA*).

57. Esherick, *Reform and Revolution*, 56–60; and Kupper, "Revolution in China," 83.

58. Li Weiyang, "Anyuan meikuang zuizao de baogong ziliao," *Jindaishi ziliao* 1 (1958), 1; and Yan Bing, "Heian de niandai: bu qu de douzheng," in *Lishi zhishi congshu: Anyuan lukuang gongren julebu shihua* (Nanchang: Jiangxi renmin chubanshe, 1983), 8.

59. "Pingsiang," 379.

60. "Bing-Wu qiyi: Xiaoshui cun (Mashi) baofa di" (Jiangxi sheng Pingxiang shi Shangli xiang Jinshan zheng, Unpublished Handout), 1.

61. "Pingsiang," 380; Charlton M. Lewis, *Prologue to the Chinese Revolution: The Transformation of Ideas and Institutions in Hunan Province, 1891–1907* (Cambridge: East Asian Research Center, Harvard University Press, 1976), 186.

62. Esherick, *Reform and Revolution*, 60.

63. Jerome Ch'en, *The Highlanders of Central China: A History, 1895–1937* (Armonk: M. E. Sharpe, 1992), 168–69. For more on the ever-normal granaries and famine relief, see Pierre-Étienne Will and R. Bin Wong, eds., *Nourish the People: The State and Civilian Granary System in China, 1650–1850* (Ann Arbor: University of Michigan Press, 1991).

64. Fu, *Liling xiangtu zhi*, 41–43.

65. *HDA*, 77; and Yan, "Hei an de nian dai," 8–9.

66. Liu, *Zhaoping zhilue*, 1234.

67. Liu, *Zhaoping zhilue*, 2291; and Esherick, *Reform and Revolution*, 60.

68. Ch'en, *Highlanders of Central China*, 25. Pingxiang County had granaries dating back at least to the Southern Song dynasty. By the founding of the Ming, five granaries were established in the four corners of the county and in the county seat. Qing-era construction included at least one ever-normal granary in the eighteenth century and several granaries after the devastation of the Taiping Rebellion in the mid-nineteenth century. See Xi Rong, ed., *Pingxiang xianzhi* (Taibei: Chengwen chubanshe, 1975), 340–41; Liu, *Zhaoping zhilue*, 855–67; and Pingxiang shi liangshi ju, eds., *Pingxiang shi liangshi zhi* (Nanchang: Jiangxi renmin chubanshe, 1992), 73 (hereafter referred to as *PXSLS*).

69. Liu, *Zhaoping zhilue*, 2290–91.

70. Yan, "Hei an de nian dai, 8–9.

71. *HDA*, 77–78; "Pingsiang," 380.

72. "Pingsiang," 380; Li, "Anyuan meikuang," 5.

73. Zhongguo kexueyuan jindaishi yanjiusuo shiliao bianyizu, eds., *Jindai shi ziliao* 1 (1958), 65 (hereafter referred to as *JDSZL*).

74. Esherick, *Reform and Revolution*, 61; Kupper, "Revolution in China," 83.

75. *HDA*, 77; and Esherick, *Reform and Revolution*, 61.

76. Lewis, *Prologue to the Chinese Revolution*, 187; and "Bing-Wu qiyi," 3.

77. Kupper, "Revolution in China," 70–78.

78. "Bing-Wu qiyi," 3–4.

79. Lewis, *Prologue to the Chinese Revolution*, 169, 187; *JDSZL*, 58; Kupper, "Revolution in China," 79–80; and Hsueh Chun-tu, *Huang Hsing and the Chinese Revolution* (Stanford: Stanford University Press, 1961), 60–61.

80. Kupper, "Revolution in China," 89–90.

81. While each of these arguments for rebellion had the ring of truth for the miners and their commoner allies, the specific spark that brought the people to violence is unclear. Some modern Chinese scholars have tried to show that this uprising was a precursor to the Sun Yatsen revolution and indicated a strong support for his Three People's Principles. Others, however, are less sanguine, pointing

out that most of the rebels were too politically unsophisticated to understand Sun's message. There is, in fact, some doubt as to whether they heard of Sun's ideas it at all. Kupper, for one, split the difference, pointing out that the ideological leanings of the leaders in Pingxiang County were more in tune with Sun's message, while the Hunanese provincial leadership supported a more pragmatic approach to organization of the commoners by emphasizing the premodern arguments for violence against the Qing. In any case, organization throughout the countryside in the three counties expanded quickly in the months leading up to the uprising. See Rao Huaimin, "Ping Liu Li qiyi shiliao de zhen wei wenti: yu 'Ping Liu Li huitang qiyi xi wen ban wei' yi wen shan qiao," *Lishi yanjiu* 6 (1995): 96–109; Wang Xuezhuang and Zhou Qiuguang, "Ping Liu Li hui tang qiyi wen ban wei: Jian lun yi ming 'Wei Zongchuan chuan,"' *Lishi yenjiu* 5 (1989): 137–52; Kupper, "Revolution in China," 88–93.

82. Lewis, *Prologue to the Chinese Revolution*, 189.

83. Pingxiang shi zheng xie, Liuyang xian zheng xie, and Liling shi zheng xie, eds., *Ping-Liu-Li qiyi ziliao* (Changsha: Hunan renmin chubanshe, 1986), 80–83 (hereafter referred to as *PLLQY*).

84. Kupper, "Revolution in China," 84.

85. *JDSZL*, 65; Yan, "Hei an de nian dai," 5.

86. Lewis, *Prologue to the Chinese Revolution*, 190; Liu, *Zhaoping zhilue*, 1131; Chang, *Rise of the Chinese Communist Party*, 8; Kupper, "Revolution in China," 85; and *PLLQY*, 183.

87. Chang, *Rise of the Chinese Communist Party*, 2.

88. Chang, *Rise of the Chinese Communist Party*, 2–3.

89. *JDSZL*, 59–61.

90. *JDSZL*, 59–61; and Chang, *Rise of the Chinese Communist Party*, 5.

91. *HDA*, 77–79; and Kupper, "Rise of the Chinese Revolution," 91.

92. *PLLQY*, 80–83, 183.

93. *PLLQY*, 183, 190, 196.

94. "Pingsiang," 380; and Lewis, *Prologue to the Chinese Revolution*, 118.

95. *HDA*, 78–79; and *JDSZL*, 58.

96. *HDA*, 78.

97. *JDSZL*, 57; and Kupper, "Revolution in China," 79.

98. Liu, *Zhaoping zhilue*, 990, 1131–33; and Kupper, "Revolution in China," 86.

99. "Pingsiang," 380; and *XHGM*, 480.

100. *PLLQY*, 183.

101. Liu, *Zhaoping zhilue*, 1549–51.

102. "Pingsiang," 380.

103. Liu, *Zhaoping zhilue*, 1133.

104. *HDA*, 79.

105. *JDSZL*, 63; and *XHGM*, 480.

106. Chang, *Rise of the Chinese Communist Party*, 4.

107. Liu, *Zhaoping zhilue*, 1133.

108. "Pingsiang," 380.
109. "Pingsiang," 380.
110. "Pingsiang," 380.
111. Liu, *Zhaoping zhilue*, 899.
112. L. C. Arlington, *Through the Dragon's Eyes: Fifty Years' Experiences of a Foreigner in the Chinese Government Services* (New York: Richard R. Smith, 1931), 196.
113. "Pingsiang," 380.
114. The years immediately after the period discussed in this book, including the rise of the Communist movement led by Mao Zedong, Liu Shaoqi, and Li Lisan, are described in great detail in Perry, *Anyuan*.

Conclusion. Industrialization in the World's Countrysides

1. G. Leinung, "An-hui Hui-t'ung mei k'uang kung ssu kai luëh = Reports on Wei-tung Coal Mining Company Ltd. Anhwei" (China: Unpublished document, 1920), 16, 31.
2. Tim Wright, *Coal Mining in China's Economy and Society, 1895–1937* (Cambridge: Cambridge University Press, 1984), 184.
3. Nym Wales, *The Chinese Communists: Sketches and Autobiographies of the Old Guard* (Westport: Greenwood Publishing Company, 1972), 94.
4. For more on this, see Changsha shi geming jinian di bagong tai and Anyuan lukuang gongren yundong jinian guang, eds., *Anyuan lukuang gongren yundong shiliao* (Changsha: Hunan gongren chubanshe, 1980), 514 (hereafter referred to as *AYLK*); and Chang Kuo-t'ao, *The Rise of the Chinese Communist Party, 1921–1927: The Autobiography of Chang Kuo-t'ao* (Lawrence: University of Kansas, Press, 1971), 16.
5. For more specific information on new cropping strategies and landlord-tenant relations, see Liu Hongpi, ed., *Zhaoping zhilue* (Taiwan: Chengwen chubanshe, 1975), 871, 887; John Lossing Buck, *Land Utilization in China: An Atlas* (Shanghai: University of Nanking, 1937), 82–83, 117; and Joseph W. Esherick, "Number Games: A Note on Land Distribution in Prerevolutionary China," *Modern China* 7, 4 (1981): 387–411.
6. Zhongguo shehui kexueyuan jindaishi yanjiusuo and Anyuan gongren yundong jinianyuan, eds., *Liu Shaoqi yu Anyuan gongren yundong* (Beijing: Zhongguo jinian yuanxue, 1981), 13 (hereafter referred to as *LSQA*).
7. Liu, *Zhaoping zhilue*, 1160–1250. Both the Xie and Zhen lineages migrated to Pingxiang County during the Ming-Qing transition during the time when virtually all the powerful families arrived. However, only one Xie lineage member received a senior licentiate degree during the Qing dynasty, and the gazetteer makes no mention of Xie Lanfang. The possibility exists that the merchant representative Xie is a member of the once-prominent lineage. Similarly, the Zhen family name is not a prominent one in the late Qing. Several Zhen family

members were members of the gentry. Also, the Zhen family was not mentioned in other discussions regarding the coalmine, and so their status in the community was likely to be modest.

8. Elizabeth J. Perry, *Anyuan: Mining China's Revolutionary Tradition* (Berkeley: University of California Press, 2012), 53, 72.

9. It is worth noting here that Wen Tingshi and members of the Li lineage are today held up as great leaders of the past. Local publications reprint some of Wen Tingshi's poetry and even declare him to be one of the founders of the Coalmines. Li Youru's brothers, Li Youtang and Li Youfen, are commemorated as successful government officials and impressive scholars. Their opposition to the court's mining policies that led to the mechanization of their county is not mentioned. See Xiao Ping and Liu Zhiyong, *Shi hua Pingxiang* (Beijing: Zhongguo wenhua chubanshe, 2006), 56, 81–84. See also, Pingxiang wenshi ziliao (7th division) and Pingxiang shi zhi tong xun (14th edition), eds., *Pingxiang renwu ji lue* (Pingxiang: Pingxiang shi zheng xie wenshi ziliao yanjiu weiyuan hui and Pingxiang shi zhi bian luo weiyuan hui, 1987), 163, 172–73.

10. Wales, *The Chinese Communists*, 85; Roy Hofheinz Jr., "The Autumn Harvest Insurrection," *The China Quarterly* 32 (1967): 46–51.

11. See especially Linda Cooke Johnson, *Shanghai: From Market Town to Treaty Port, 1074–1858* (Stanford: Stanford University Press, 1995). Some books that focus on the last decades of the nineteenth century and the early twentieth century provide evidence of long-term gradual development prior to modernization. See, for example, David Buck, *Urban Change in China: Politics and Development in Tsinan, Shantung, 1890–1949* (Madison: University of Wisconsin Press, 1978).

Glossary

Anyuan 安源
Aozhou shuyuan 鳌州書院
ba 把 (also written as 巴)
Bafangjin 八方井
Bai Lizhi 柏李治
bangban 帮辦
bangdong 帮董
bao 保
Bao he gong zhuang 保 合公庄
Beitang gongci 北堂公祠
Cai Shaonan 蔡紹南
chi 尺
Ci Fei 雌飛
cun 寸
da ju 大舉
Daye 大冶
Deng 邓
dian 典
Duban 督辦
dumai 杜賣
Duan Xin 段鑫
Gan Chengqing 甘成清
Gaokeng 高坑
Gelaohui 歌老会
Gong Chuntai 龔春臺
Gu Jiaxiang 顧家相
guandu shangban 官督商辦
Guangtaifu 廣泰福
Hakka 客家
Hongzihao 洪字号
Huguang 湖廣

Hongjianghui 洪江会
Hui Jixun 辉積勳
jiao 角
Jinggang Mountains 井岗山
Jinyushi 金鱼石
juemai 絶賣
Leiyang County 耒陽縣
Li Hanzhang 李瀚章
Li Hongzhang 李鴻章
lijin 厘金
Lijin Bureau 厘金局
Li Jingshu 黎景淑
Li Liejun 李烈鈞
Li Lisan 李立三
Li Shaonan 李烧南
Li Shaobai 李少白
Li Youfen 李有芬
Li Youru 李有架
Li Youtang 李有棠
Li Youxiang 李有薌
Li Yu 李豫
Li Zongjin 李宗琎
Liling City 醴陵城
Liling County 醴陵縣
Lin Zhixi 林志熙
Liu Kunyi 劉坤一
Liuyang County 瀏陽縣
Lu Hongchang 盧洪昶
Lu River 綠江
luan 鑾
Lukou 綠口
Luoxiao Mountain range 羅霄山脈
Luxi 蘆溪
Ma'anshan 馬鞍山
Ma Fuyi 馬福益
mu 亩
Ouyang Bingrong 歐陽柄榮
Peng 鵬
Pengmin 棚民
Ping River 萍江

Pingxiang City 萍鄉城
Pingxiang County 萍鄉縣
po 岥
shan zhu 山主
shang hao 商号
Shangli 上栗市
Sheng Xuanhuai 盛宣懷
Shi Dakai 石達開
Song Zhishou 宋志寿
tang 塘
Tianzi Mountain 天子山
tidiao 提調
Tongmenghui 同盟绘
tongxiang 同鄉
tu 圖
Wangjiayuan 王甲源
weiyuan 委員
Wei Zongquan 魏宗銓
Wen Pinshan 文品山
Wen Tianyi 文天一
Wen Tingbi 文廷弼
Wen Tingjun 文廷鈞
Wen Tingshi 文廷式
Wu Shizhang 吳式璋
Xiangdong 湘東市
Xiao Kechang 蕭克昌
Xiao Liyan 蕭立炎
Xiao Ruofeng 蕭若鋒
xieban 协辦
Xu Yinhui 許寅輝
Yichun County 宜春縣
Yuan River 袁江
Yuanzhou Fu 袁州府
Yueni 粵逆
Yun Jixun 惲積勛
zhang 丈
zhang 嶂
Zhang Guotao 張國燾
Zhang Zanchen 張贊宸
Zhang Zhirui 張之鋭

Zheng Guanying 鄭觀應
Zijiachong 紫家冲
zong kuangshi 總鑛师
zongban 總辦

Bibliography

Ali, Saleem H. *Mining, the Environment, and Indigenous Development Conflicts*. Tucson: University of Arizona Press, 2003.

Arlington, L. C. *Through the Dragon's Eyes: Fifty Years' Experiences of a Foreigner in the Chinese Government Services*. New York: Richard R. Smith, 1931.

Arrigo, Linda Gale. "Landownership Concentration in China: The Buck Survey Revisited." *Modern China* 12, 3 (1986): 259–360.

Atkinson, G. Anthony. "British Architects in the Tropics." *Architectural Association Journal* 69 (1953): 7–21.

Averill, Stephen Carl. "Revolution in the Highlands: The Rise of the Communist Movement in Jiangxi Province." PhD dissertation. Cornell University, 1982.

———. *Revolution in the Highlands: China's Jinggangshan Base Area*. Lanham: Rowman & Littlefield, 2006.

Ballard, Walter J. "As Our Consultants Say." *Los Angeles Times* (9 April 1908): 115.

Bays, Daniel H. "The Nature of Provincial Political Authority in Late Ch'ing Times: Chang Chih-tung in Canton, 1884–1889." *Modern Asian Studies* 4, 4 (1970): 325–47.

Berkowitz, N. *An Introduction to Coal Technology*. New York: Academic Press, 1979.

Bernhardt, Kathryn. *Rents, Taxes, and Peasant Resistance: The Lower Yangzi Region, 1840–1950*. Stanford: Stanford University Press, 1992.

Bickmore, Albert S. "Sketch of a Journey from Canton to Hankow." *Journal of the Royal Geographical Society of London* 38 (1868): 50–68.

Billingsley, Phil. "Bandits, Bosses, and Bare Sticks: Beneath the Surface of Local Control in Early Republican China." *Modern China* 7, 3 (1981): 235–88.

"Bing-Wu qiyi: Xiaoshui cun (Mashi) baofa di" (The 1906 Uprising: Xiaoshui Village, the Place Where It Erupted). Unpublished handout: Jiangxi sheng Pingxiang shi Shangli xiang Jinshan zheng.

Bonnell, Victoria E. *Roots of Rebellion: Worker's Politics and Organization in St. Petersburg and Moscow, 1900–1914*. Berkeley: University of California Press, 1983.

Braisted, William R. "The United States and the American China Development Company." *Far Eastern Quarterly* 11, 2 (1952): 147–65.

Bray, Francesca. *Science and Civilization in China, Volume 6: Biology and Biological Technology, Part II: Agriculture*. Cambridge: Cambridge University Press, 1984.

———. *Rice Economies: Technology and Development in Asian Societies*. Oxford: Basil Blackwell, 1986.

———. *Technology and Gender: Fabrics of Power in Late Imperial China*. Berkeley: University of California Press, 1997.

Brenner, Robert. "Agrarian Class Structure and Economic Development in Pre-Industrial Europe." *Past and Present* 70 (1976): 30–75.

———. "The Origins of Capitalist Development: A Critique of Neo-Smithian Marxism." *New Left Review* 104 (1977): 25–92.

———. "The Agrarian Roots of European Capitalism." *Past and Present* 97 (1982): 16–113.

———."The Social Basis of Economic Development." In *Analytical Marxism*, 23–53. Ed. John Roemer. Cambridge: Cambridge University Press, 1986.

Brenner, Robert, and Christopher Isett. "England's Divergence from China's Yangzi Delta: Property Relations, Microeconomics, and Patterns of Development." *Journal of Asian Studies* 61, 2 (2002): 609–62.

Brook, Timothy. "Family Continuity and Cultural Hegemony: The Gentry of Ningbo, 1368–1911." In *Chinese Local Elites and Patterns of Dominance*, 27–50. Ed. Joseph W. Esherick and Mary Backus Rankin. Berkeley: University of California Press, 1990.

Brown Shannon R., and Tim Wright. "Technology, Economics, and Politics in the Modernization of China's Coal-Mining Industry, 1850–1895." *Explorations in Economic History* 18, 1 (1981): 60–83.

Buck, David. *Urban Change in China: Politics and Development in Tsinan, Shantung, 1890–1949*. Madison: University of Wisconsin Press, 1978.

Buck, John Lossing. *Land Utilization in China: An Atlas*. Shanghai: University of Nanking, 1937.

Buoye, Thomas M. "Litigation, Legitimacy, and Lethal Violence: Why County Courts Failed to Prevent Violent Disputes over Property in Eighteenth-Century China." In *Contract and Property in Early Modern China*, 94–119. Ed. Madeline Zelin, Jonathan Ocko, and Robert Gardella. Stanford: Stanford University Press, 2004.

Campregher, Christoph. "Shifting Perspectives on Development: An Actor-Network Study of a Dam in Costa Rica." *Anthropological Quarterly* 83, 4 (2010): 783–804.

Carlson, Ellsworth C. *The Kaiping Mines, 1877–1912*. Cambridge: Harvard University Press, 1971.

Chakrabarty, Dipesh. "Conditions for Knowledge of Working-Class Conditions: Employers, Government, and the Jute Workers of Calcutta, 1890–1940." In *Selected Subaltern Studies*, 179–230. Ed. Ranajit Guha and Gayatri Chakravorty Spivak. New York: Oxford University Press, 1988.

Chan, Wellington K. K. *Merchants, Mandarins, and Modern Enterprise in Late Ch'ing China*. Cambridge: Harvard University Press, 1975.

———. *Politics and Industrialization in Late Imperial China*. Singapore: Institute of Southeast Asian Studies, 1975.

Chandler, Alfred D. *The Visible Hand: The Managerial Revolution in American Business*. Cambridge: Harvard University Press, 1977.

———. "The Emergence of Managerial Capitalism." *Business History Review* 58, 4 (1984): 473–503.

Chang Kuo-t'ao. *The Rise of the Chinese Communist Party, 1921–1927: The Autobiography of Chang Kuo-t'ao*. Lawrence: University of Kansas Press, 1971.

Chang Sen-dou. "Some Aspects of the Urban Geography of the Chinese Hsien Capital." *Annals of the Association of American Geographers* 51, 1 (1951): 23–45.

———. "The Morphology of Walled Capitals." In *The City in Late Imperial China*, 75–100. Ed. G. William Skinner. Stanford: Stanford University Press, 1977.

Chao, Kang. "New Data on Land Ownership Patterns in Ming-Ch'ing China: A Research Note." *Journal of Asian Studies* 40, 4 (1981): 719–34.

———. *Man and Land in Chinese History: An Economic Analysis*. Stanford: Stanford University Press, 1986.

Chayanov, A. V. *The Theory of Peasant Economy*. Madison: University of Wisconsin Press, 1986.

Chen, fu-mei Chang, and Ramon Myers. "Customary Law and the Economic Growth of China during the Ch'ing Period," Part 2. *Ch'ing-shih wen-t'i* 3, 10 (1976): 4–27.

Ch'en, Jerome. *The Highlanders of Central China: A History, 1895–1937*. Armonk: M. E. Sharpe, 1992.

Chen Xulu and Gu Tinglong, eds. *Hanyeping gongsi* (Hanyeping Company), Vol. 1. Shanghai: Shanghai renmin chubanshe, 1984.

Chen Xulu and Gu Tinglong, eds. *Hanyeping gongsi* (Hanyeping Company), Vol. 2. Shanghai: Shanghai renmin chubanshe, 1986.

Chen Zhen. *Zhongguo jindai gongyeshi ziliao* (Historical materials of China's modern industry), Vol. 3. Beijing: N.p., 1961.

Chesneaux, Jean. *The Chinese Labor Movement, 1919–1927*. Trans. H. M. Wright. Stanford: Stanford University Press, 1968.

Chisholm, George G. "The Resources and Means of Communication of China." *The Geographical Journal* 12, 5 (1898): 500–19.

Ch'u Tung-tsu. *Local Government in China under the Ch'ing*. Stanford: Stanford University Press, 1962.

Cohen, Myron L. "Writs of Passage in Late Imperial China: The Documentation of Practical Understandings in Minong, Taiwan." In *Contract and Property in Early Modern China*, 37–93. Ed. Madeline Zelin, Jonathan Ocko, and Robert Gardella. Stanford: Stanford University Press, 2004.

Collins, William F. *Mineral Enterprise in China*. New York: Macmillan, 1918.

Cooper, Frederick. "Urban Space, Industrial Time, and Wage Labor in Africa." In *Struggle for the City: Migrant Labor, Capital, and the State in Urban Africa*, 7–50. Ed. Frederick Cooper. Beverley Hills: Sage, 1983.

———. *Colonialism in Question: Theory, Knowledge, History*. Berkeley: University of California Press, 2005.

Davis, Fei-ling. *Primitive Revolutionaries of China: A Study of Secret Societies of the Late Nineteenth Century*. Honolulu: University of Hawaii Press, 1971.

Deyo, Frederic C. "Labor and Development Policy in East Asia." *Annals of the American Academy of Political and Social Science* 505 The Pacific Region: Challenges to Policy and Theory (1989): 152–61.

Dickinson, Frederick R. *War and National Reinvention: Japan in the Great War, 1914–1919*. Cambridge: Harvard University Press, 1999.

Duara, Prasenjit. *Culture, Power, and the State: Rural North China, 1900–1942*. Stanford: Stanford University Press, 1988.

Duus, Peter, Ramon Hawley Myers, and Mark R. Peattie, eds. *The Japanese Informal Empire in China, 1895–1937*. Princeton: Princeton University Press, 1989.

Elvin, Mark. *The Pattern of the Chinese Past*. Stanford: Stanford University Press, 1973.

———. "The Administration of Shanghai, 1905–1914." In *The Chinese City between Two Worlds*, 239–62. Ed. Mark Elvin and G. William Skinner. Stanford: Stanford University Press, 1974.

Encyclopedia Britannica: A Dictionary of Arts, Sciences and General Literature. 9th edition. New York: Charles Scribner's Sons, Various Dates.

Esherick, Joseph W. *Reform and Revolution in China: The 1911 Revolution in Hunan and Hubei*. Berkeley: University of California Press, 1976.

———. "Number Games: A Note on Land Distribution in Prerevolutionary China." *Modern China* 7, 4 (1981): 387–411.

Esthus, Raymond A. "The Changing Concept of the Open Door, 1899–1910." *The Mississippi Valley Historical Review* 46, 3 (1959): 435–54.

Evens, Francis T. "Roads, Railways, and Canals: Technical Choices in 19th-Century Britain." In *Technology and the West: A Historical Anthology from "Technology and Culture,"* 199–232. Ed. Terry S. Reynolds and Stephen H. Cutcliffe. Chicago: University of Chicago Press, 1997.

Fang Zhuofen, Hu Tiewen, Juan Rui, and Fang Xing. "Capitalism during the Early and Middle Qing." In *Chinese Capitalism, 1522–1840*, 249–371. Ed. Xu Dixin and Wu Chengming. Ed. and annotated by C. A. Curwen. Trans. Li Zhengse, Liang Mioru, Li Siping. New York: St. Martin's Press, 2000.

Faure, David. *The Structure of Chinese Rural Society: Lineage and Village in the Eastern New Territories, Hong Kong*. Hong Kong: Oxford University Press, 1986.

Feuerwerker, Albert. "China's Nineteenth-Century Industrialization: The Case of the Hanyeping Coal and Iron Company, Limited." In *The Economic Development of China and Japan: Studies in Economic History and Political Economy*, 79–110. Ed. C. D. Cowan. New York: Frederick A. Praeger, 1964.

———. *China's Early Industrialization: Sheng Hsuan-huai (1844–1916) and Mandarin Enterprise*. New York: Antheneum, 1970.

———. "Economic Trends in the Late Ch'ing Empire, 1870–1911." In *The Cambridge History of China, Volume 11 Part 2: Late Ch'ing, 1800–1911*, 1–69.

Ed. Denis Twitchett and John King Fairbank. Cambridge: Cambridge University Press, 1980.

Freedman, Maurice. *Lineage Organization in Southeastern China*. New York: Humanities Press, 1965.

Fu Xiongxiang, ed. *Liling xiangtu zhi* (Liling County local gazetteer). Taiwan: Chengwen chubanshe, 1926.

Fukuyama, Francis. "The Imperative of State Building." *Journal of Democracy* 15, 2 (2004): 17–31.

Fukuyama, Francis, and Sanjay Marwah. "Dimensions of Development." *Journal of Democracy* 11, 4 (2000): 80–94.

Gamble, Sidney. "Daily Wages of Unskilled Chinese Laborers, 1807–1902." *Far East Quarterly* 3, 1 (1943): 41–73.

Gates, Hill. "Dependency and the Part-Time Proletariat in Taiwan." *Modern China* 5, 3 (1979): 381–407.

———. *China's Motor: A Thousand Years of Petty Capitalism*. Ithaca: Cornell University Press, 1996.

Gaubatz, Piper Rae. *Beyond the Great Wall: Urban Form and Transformation on the Chinese Frontiers*. Stanford: Stanford University Press, 1996.

Geertz, Clifford. *Agricultural Involution: The Processes of Ecological Change in Indonesia*. Berkeley: University of California Press, 1963.

Gereffi, Gary. "Development Strategies and the Global Factory." *Annals of the American Academy of Political and Social Science* 505, The Pacific Region: Challenges to Policy and Theory (1989): 92–104.

Gerschenkron, Alexander. *Economic Backwardness in Historical Perspective: A Book of Essays*. Cambridge: Harvard University Press, 1962.

Golas, Peter J. *Science and Civilization in China: Volume 5: Chemistry and Chemical Technology, Part XII: Mining*. Cambridge: Cambridge University Press, 1999.

Grosevernor, W. Clayton. "The Province of Hunan: Some Characteristics and Peculiarities." *Scottish Geographical Magazine* 44 (1928): 144–50.

Gu Jiaxiang. Chouban Pingxiang tielu gong du (Public documents on the preparation of the Pingxiang Railroad). Pingxiang: Pingxiang xian chubanshe, 1900.

———. *Lutang wenji* (Gu Jiaxiang's collected works). Taibei: Wenhai chubanshe, 1972.

Gu Lang. *Zhongguo shi da kuang chang diaocha ji* (Investigations of China's ten largest mines). Shanghai: Commercial Press, 1916.

Guo shi guan eds. *Zhongguo tielu yange shi* (History of iron ore mining in China). Taibei: Taibei xian xindian shi, 1984.

Haggard, Stephan, and Steven B. Webb. "What Do We Know about the Political Economy of Economic Policy Reform?" *The World Bank Research Observer* 8, 2 (1993): 143–68.

Hansen, Valerie. *Negotiating Daily Life in Traditional China: How Ordinary People Used Contracts, 600–1400*. New Haven: Yale University Press, 1995.

Hanwell, Norman D. "Tayeh Iron Fields Put under Direct Japanese Control." *Far Eastern Survey* 8, 4 (1939): 47.

Hartwell, Robert. "A Cycle of Economic Change in Imperial China: Coal and Iron in Northeast China, 750–1350." *Journal of the Economic and Social History of the Orient* 10 (1967): 102–55.

Headrick, Daniel R. *The Tools of Empire: Technology and European Imperialism in the Nineteenth Century*. New York: Oxford University Press, 1981.

Hershatter, Gail. *The Workers of Tianjin, 1900–1949*. Stanford: Stanford University Press, 1986.

———. "The Hierarchy of Shanghai Prostitution, 1870–1949." *Modern China* 15, 4 (1989): 463–98.

Hevia, James L. *English Lessons: The Pedagogy of Imperialism in Nineteenth-Century China*. Durham: Duke University Press, 2003.

Ho Ping-ti. *Studies on the Population of China, 1368–1953*. Cambridge: Harvard University Press, 1959.

Hoag, Heather J., and May-Britt Öhman. "Turning Water into Power: Debates over the Development of Tanzania's Rufiji River Basin, 1945–1985." *Technology and Culture* 49, 3 (2008): 624–51.

Hoang, Peter. "A Practical Treatise on Legal Ownership." *Journal of the China Branch of the Royal Asiatic Society of Great Britain and Ireland* 23 (1889): 118–74.

Hobsbawm, Eric. *Bandits*. Revised edition. New York: Random House, 1981.

Hofheinz, Roy, Jr. "The Autumn Harvest Insurrection." *The China Quarterly* 32 (1967): 37–87.

———. "The Ecology of Chinese Communist Success: Rural Influence Patterns, 1923–45." In *Chinese Communist Politics in Action*, 3–77. Ed. A. Doak Barnett. Seattle: University of Washington Press, 1969.

Hommel, Rudolf P. *China at Work: An Illustrated Record of the Primitive Industries of China's Masses, Whose Life Is Toil, and Thus an Account of Chinese Civilization*. New York: John Day Company, 1937.

"Hookworm Disease in China." *British Medical Journal* 1, 3460 (1927): 807.

Hoover, H. C. "Present Situation of the Mining Industry in China." *Engineering and Mining Journal* 69 (1900): 619–20.

Horetz, Peter J. "China's Hookworms." *The China Quarterly* 172 (2002): 1029–41.

Hornibrook, Jeff. "Local Elites and Mechanized Mining in China: The Case of the Wen Lineage in Pingxiang County, Jiangxi." *Modern China* 27, 2 (2001): 202–28.

———. "Riding the Tiger: Merchant-State Alliance in a Coalmine Modernisation Scheme." *Business History* 45, 2 (2003): 35–51.

———. "Don't Tread on Me: Land, Officials, and Archival Work on a Qing Dynasty Mining Enterprise." *Chinese Business History* 15, 1 (2005): 1–2, 9.

Hou, T. F. *General Statement on the Mining Industry*. Vol. 3. Beijing: Ministry of Agriculture and Mines, 1929.

———. *General Statement on the Mining Industry*. Vol. 5. Beijing: Geological Survey of China, 1935.

Hoyt, Lansing W. "Blast Furnaces and Steel Mills in China: A Comprehensive Survey of China's Steel Industry." *Far Eastern Review* 19, 5 (1923): 305–20.

Hsiao, C. C. "Criteria for the Evaluation of Blast-Furnace Coke." *Geological Bulletin* 30 (1937): 77–87.

Hsiao Kung-chuan. *Rural China: Imperial Control in the Nineteenth-Century.* Seattle: University of Washington Press, 1960.

Hsueh Chun-tu. *Huang Hsing and the Chinese Revolution.* Stanford: Stanford University Press, 1961.

Hsueh, K. L. "The Iron and Steel Industry of China." *The Far Eastern Review* 24, 4 (1928): 176–83.

———. "The Iron and Steel Industry of China." *Chinese Economic Journal* 2, 1 (1928): 1–25.

Hua Wen and Luo Xiao. "Pingxiang meitan fazhan shilue." In *Pingxiang meitan fazhan shilue*, 1–15. Ed. Jiangxi sheng zheng xi wenshi ziliao yanjiu weiyuan hui and Pingxiang shi zheng wei wenshi ziliao yanjiu weiyuan hui he bian. Hong Kong: N.p., 1987.

Huang Liu-hung. *A Complete Book Concerning Happiness and Benevolence: A Manual for Local Magistrates in Seventeenth-Century China.* Trans. and ed. Djan Chu. Tucson: University of Arizona Press, 1984.

Huang, Philip C. C. *The Peasant Economy and Social Change in North China.* Stanford: Stanford University Press, 1985.

———. *The Peasant Family and Rural Development in the Yangzi Delta, 1350–1988.* Stanford: Stanford University Press, 1990.

———. "The Paradigmatic Crisis in Chinese Studies: Paradoxes in Social and Economic History." *Modern China* 17, 3 (1991): 299–341.

———. *Code, Custom, and Legal Practice in China: The Qing and Republic Compared.* Stanford: Stanford University Press, 2001.

———. "Development or Involution in Eighteenth-Century Britain and China? A Review of Kenneth Pomeranz's The Great Divergence: China, Europe, and the Making of Modern World Economy." *Journal of Asian Studies* 61, 2 (2002): 501–38.

———. "Further Thoughts on Eighteenth-Century Britain and China: Rejoinder to Pomeranz's Response to My Critique." *Journal of Asian Studies* 62, 1 (2003): 157–67.

Huang, Ray. "The Lung-ch'ing and Wan-li Reigns, 1567–1620." In *The Cambridge History of China: Volume 7: The Ming Dynasty, 1368–1644, Part I*, 511–84. Ed. Frederick W. Mote and Denis Twitchett. Cambridge: Cambridge University Press, 1988._

Huang Shiguo, ed. *Pingxiang shi zhi* (Pingxiang city gazetteer). Pingxiang: Pingxiang shi zhi bianzuan weiyuan huibian gangzhi chubanshe chuban, 1996.

Huang Zuxun, ed. *Liuyang xiangtu zhi* (Liuyang county local gazetteer). Taibei: Qingshui yingshua chubanshe, 1967.

Huggins, H. C. "Japan's Iron and Steel Industry." *Far Eastern Review* 19, 3 (1923): 190–98.

Hummel, Arthur W., ed. *Eminent Chinese of the Ch'ing Period (1644–1912)*. Washington: United States Government Printing Office, 1943.

"Investigation Report." In *Chinese Civilization and Society: A Sourcebook*, 233–34. Ed. Patricia Buckley Ebrey. New York: The Free Press, 1981.

Isaacman, Allen, and Barbara Isaacman. "Resistance and Collaboration in Southern and Central Africa, ca. 1850–1920." *International Journal of African Historical Studies* 10, 1 (1977): 31–62.

Isaacman, Allen, and Chris Sneddon. "Toward a Social and Environmental History of the Building of Cahora Bassa Dam." *Journal of Southern African Studies* 26, 4 (2000): 597–632.

Isett, Christopher Mills. *State, Peasant, and Merchant in Qing Manchuria, 1644–1862*. Stanford: Stanford University Press, 2007.

Jamieson, George. *Chinese Family and Commercial Law*. Hong Kong: Vetch and Lee, 1970.

Jansen, Mauris B. "Yawata, Hanyehping, and the Twenty-One Demands." *Pacific Historical Review* 23, 1 (1954): 31–48.

Johnson, Linda Cooke. *Shanghai: From Market Town to Treaty Port, 1074–1858*. Stanford: Stanford University Press, 1995.

Johnson, William Reid. "Hanyang Iron Works, 1890–1908: A Key Enterprise in China's Industrialization." MA thesis. University of Washington, 1955.

Jones, William C. *The Great Qing Code*. Oxford: Clarendon Press, 1994.

Kao, P., and H. C. Hsu. "Geology of Western Kiangsi." *Geological Memoirs* 16 (1940): 61–72.

Kennedy, Thomas L. "Chang Chih-tung and the Struggle for Strategic Industrialization: The Establishment of the Hanyang Arsenal, 1884–1895." *Harvard Journal of Asiatic Studies* 30 (1973): 154–82.

Kent, Percy Horace. *Railway Enterprise in China: An Account of Its Origin and Development*. London: Edward Arnold, 1907.

Kerr, Rose, and Nigel Wood. *Science and Civilisation in China: Volume 5: Chemistry and Chemical Technology, Part XII: Ceramic Technology*. Ed. Rose Kerr. With contributions from Ts'ai Mei-fen and Zhang Fukang. Cambridge: Cambridge University Press, 2004.

King, Anthony D. *The Bungalow: The Production of a Global Culture*. London: Routledge and Kegan Paul, 1984.

Kleeman, Terry F. "Land Contracts and Related Documents." In *Chugoku no shukyo shiso to kagaku: Makio Ryokai Hakushi shoju kinen ronshu*, 1–34. Ed. Makio Ryokai. Tokyo: Kokusho Kankohkai, 1984.

Köll, Elisabeth. "Recent Debates in the Field of Business History: What They Mean for China Historians." *Chinese Business History* 10, 1 (2000): 1–2.

———. *From Cotton Mill to Business Empire: The Emergence of Regional Enterprises in Modern China*. Cambridge: Harvard University Press, 2003.

Kuhn, Philip A. *Rebellion and Its Enemies in Late Imperial China: Militarization and Social Structure, 1796–1864*. Cambridge: Harvard University Press, 1980.

Kupper, Samuel Yale. "Revolution in China: Kiangsi Province, 1905–1913." PhD dissertation. University of Michigan, 1973.

Kwan Man Bun. "Mapping the Hinterland: Treaty Ports and Regional Analysis in Modern China." In *Remapping China: Fissures in Historical Terrain*, 181–93. Ed. Gail Hershatter, Emily Honig, Jonathan Lippman, and Randall Stross. Stanford: Stanford University Press, 1996.

Kwong, Luke S. K. "The T'i-Yung Dichotomy and the Search for Talent in Late-Ch'ing China." *Modern Asian Studies* 27, 2 (1993): 253–79.

Lai, Chi-kong. "Li Hong-chang and Modern Enterprise: The China Merchants' Company, 1872–1885." In *Li Hung-chang and China's Early Modernization*, 216–47. Ed. Samuel C. Chu and Kwang-Ching Liu. Armonk: M. E. Sharpe, 1994.

Landes, David S. *The Unbound Prometheus: Technological Change and Industrial Development in Western Europe from 1750 to the Present*. Cambridge: Cambridge University Press, 1969.

Lee, En-han. "China's Response to Foreign Investment in Her Mining Industry." *Journal of Asian Studies* 28, 1 (1968): 55–76.

———. "China's Response to the Foreign Scramble for Railway Concessions, 1895–1911." *Journal of Oriental Studies* 14, 1 (1976): 1–22.

Lee, V. K. "China's Iron and Steel Industry." *The Chinese Students Monthly* 9, 4 (1914): 284–90.

Leinung, G. "An-hui Hui-t'ung mei k'uang kung ssu kai lüeh = Report on Wei-tung Coal Mining Company Ltd. Anhwei." China: Unpublished document, 1920.

Leong, Sow-Theng. *Migration and Ethnicity in Chinese History: Hakkas, Pengmin, and Their Neighbors*. Ed. Tim Wright. Stanford: Stanford University Press, 1997.

Leung, Yuen-sang. *The Shanghai Taotai: Linkage Man in a Changing Society, 1843–90*. Honolulu: Asian Studies at Hawaii, 1990.

Leung, Philip Yuen-sang. "Crisis Management and Institutional Reform: The Expectant Officials in the Late Qing." In *Dragons, Tigers, and Dogs: Qing Crisis Management and the Boundaries of State Power in Late Imperial China*, 61–77. Ed. Robert J. Antony and Jane Kate Leonard. Ithaca: Cornell University Press, 2002.

Leutner, Mechtild, and Andreas Steen. *Deutsch-chinesische Beziehungen 1911–1927: vom Kolonialismus zur "Gleichberechtigung": eine Quellensammlung*. Berlin: Akademie Verlag, 2006.

Lewis, Charlton M. *Prologue to the Chinese Revolution: The Transformation of Ideas and Institutions in Hunan Province: 1891–1907*. Cambridge: East Asian Research Center, Harvard University Press, 1976.

Li Jui. *The Early Revolutionary Activities of Comrade Mao Tse-tung*. White Plains: M. E. Sharpe, 1997.

Li Weiyang. "Anyuan meikuang zuizao de baogong ziliao" (Materials on the Anyuan Coalmine's First Labor Strikes). *Jindaishi ziliao* 1 (1958): 1–7.

Liang, Shih-shih, editor in chief. *A New Practical Chinese-English Dictionary*. Taipei: The Far East Book Company, 1988.

Liu Hongpi, ed. *Zhaoping zhilue* (Pingxiang county gazetteer). Taiwan: Chengwen chubanshe, 1975; originally published 1935.

Liu Kunyi. *Liu Kunyi yiji* (The collected works of Liu Kunyi). Beijing: Zhonghua shuju chubanshe, 1959.

Liu Minghan, and Ma Jingyuan, eds. *Hanyeping gongsi zhi* (Documents of the Hanyeping Company). Wuhan: Huazhong li gong daxue chubanshe, 1990.

Liu, Kwang-Ching. "The Beginnings of China's Modernization." In *Li Hung-chang and China's Early Modernization*, 3–14. Ed. Samuel C. Chu and Kwang-Ching Liu. Armonk: M. E. Sharpe, 1994.

Liu Ping. *Zhang Zhidong zhuan* (Biography of Zhang Zhidong). Lanzhou: Lanzhou daxue chubanshe, 2000.

Liu Zongdao. "Lishi pengbei wandai yongcun: fang Anyuan sanji" (The historical accomplishments will last 10,000 generations: recollections on a trip to Anyuan). *Dang de shenghuo congkan* 3 (1980): 37–39.

Lufrano, Richard John. *Honorable Merchants: Commerce and Self-Cultivation in Late Imperial China*. Honolulu: University of Hawaii Press, 1997.

Luo Xiao. *Pingxiang shi difang meitan gongye zhi* (Documents of Pingxiang City's local coal production). Nanchang shi: Jiangxi renmin chubanshe, 1992.

Mann, Susan. *Local Merchants and the Chinese Bureaucracy, 1750–1950*. Stanford: Stanford University Press, 1987.

Mao Zedong. "Report on an Investigation of the Peasant Movement in Hunan." In *Selected Works of Mao Tse-tung*, Vol. 1, 23–62. Beijing: Foreign Languages Press, 1967.

———. *Report from Xunwu*. Translated with an introduction by Roger R. Thompson. Stanford: Stanford University Press, 1990.

Mathews, R. H. *Mathews' Chinese-English Dictionary*. Revised American Edition. Cambridge: Harvard University Press, 1931.

Mazumdar, Sucheta. *Sugar and Society in China: Peasants, Technology and the World Market*. Cambridge: Harvard University Press, 1998.

McDonald, Angus. *The Urban Origins of Rural Revolution: Elites and the Masses in Hunan Province, China, 1911–1927*. Berkeley: University of California Press, 1978.

Meillassioux, Claude. "From Reproduction to Production: A Marxist Approach to Economic Anthropology." *Economy and Society* 1, 1 (1972): 93–105.

Morrison, G. James. "Journeys in the Interior of China." *Proceedings of the Royal Geographical Society and Monthly Record of Geography* 2, 3 (1880): 145–66.

Nishizawa, Kimio. "Mines in the Yangtze Valley, China." *Journal of the Royal Society of Arts* 58 (1910): 935–45.

"Notes on Coal Mines in China." *The Far Eastern Review* 37 (1931): 202–04.

Osborne, Anne Rankin. "Barren Mountains, Raging Rivers: The Ecological and Social Effects of Changing Land Use in the Yangzi Periphery in Late Imperial China." PhD dissertation. Columbia University, 1989.

Parsons, William Barclay. *Report on the Survey and Prospects of a Railway between Hankow and Canton, under the Concession by the Chinese Government to the American China Development Company*. New York: American China Development Corporation, 1899.

———. *An American Engineer in China*. New York: McClure, Phillips and Company, 1900.

———. "From the Yang-Tse Kiang to the China Sea." *The Geographical Journal* 19, 6 (1902): 711–35.

———. *Railways in China*. Philadelphia: Engineers Club of Philadelphia, 1916.

Pasternak, Burton. *Kinship and Community in Two Chinese Villages*. Stanford: Stanford University Press, 1972.

Perdue, Peter. *Exhausting the Earth: State and Peasant in Hunan, 1500–1850*. Cambridge: Harvard University Press, 1987.

Perez-Pena, Richard. "Gen. Yang Dezhi, 83, Dies: Was Chinese Army Chief." *New York Times* (28 October 1994): A-15.

Perkins, Dwight. *Agricultural Development in China, 1368–1968*. Chicago: Aldine Press, 1969.

Perry, Elizabeth J. *Anyuan: Mining China's Revolutionary Tradition*. Berkeley: University of California Press, 2012.

"The Pingsiang Colliery: A Story of Early Mining Difficulties in China." *The Far Eastern Review* 12, 10 (1916): 375–80.

Ping, Wen. "Zhang Zanchen jian jie" (A short depiction of Zhang Zanchen). In *Pingxiang meitan fazhan shilue*, 194–95. Ed. Jiangxi sheng zheng xi wenshi ziliao yanjiu weiyuan hui and Pingxiang shi zheng wei wenshi ziliao yanjiu weiyuan hui he bian. Hong Kong: N.p., 1987.

Polachek, James M. "The Moral Economy of the Kiangsi Soviet (1928–1934)." *Journal of Asian Studies* 42, 4 (1983): 805–29.

Pomeranz, Kenneth. *The Making of a Hinterland: State Society and Economy in Inland North China, 1853–1937*. Berkeley: University of California Press, 1993.

———. " 'Traditional' Chinese Business Firms Revisited: Family, Firm, and Financing in the History of the Yutang Company of Jining, 1779–1956." *Late Imperial China* 18, 1 (1997): 1–38.

———. *The Great Divergence: China, Europe, and the Making of the Modern World Economy*. Princeton: Princeton University Press, 2000.

———. "Beyond the East-West Binary: Resituating Development Paths in the Eighteenth-Century World." *Journal of Asian Studies* 61, 2 (2002): 539–90.

———. "Facts Are Stubborn Things: A Response to Philip Huang." *Journal of Asian Studies* 62, 1 (2003): 167–81.

"The Present Financial Status of the Hanyehping Company, with a List of Outstanding Foreign Loans." *The Far Eastern Review* 23, 7 (1927): 329.

Price, Polly J. *Property Rights: Rights and Liberties under the Law*. Santa Barbara: ABC-CLIO, 2003.

Quan Hansheng, ed. *Hanyeping gongsi shilue* (Historical materials of the Hanyeping Company). Xianggang: Xianggang zhongwen daxue chubanshe, 1972.

Rao Huaimin. "Ping Liu Li qiyi shiliao de zhen wei wenti: yu 'Ping Liu Li huitang qiyi xi wen ban wei' yi wen shang qiao" (True or false? Problems concerning historical materials on the Pingxiang-Liuyang-Liling Uprising: In questions of "ascertaining the falsification of two uprising manifestos of secret societies in Pingxiang, Liuyang, and Liling"). *Lishi yanjiu* 6 (1995): 96–109.

Rawski, Thomas G. *Economic Growth in Prewar China*. Berkeley: University of California Press, 1989.

Rea, Geo Bronson. "Railway Loan Agreements and Their Relations to the Open Door: A Plea for Fair Play to China." *Far Eastern Review* 6, 6 (1909): 215–27.

Read, Thomas T. "The Mineral Production and Resources of China." *Transactions of the American Institute of Mining Engineers* 43 (1912): 3–53.

Reed, Alfred C. "Coal Mining in China." *The Scientific Monthly* 5 (1917): 36–49.

Reed, Bradly W. *Talons and Teeth: County Clerks and Runners in the Qing Dynasty*. Stanford: Stanford University Press, 2000.

Ren Mei'e, Yang Renzhang, and Bao Haosheng. *An Outline of China's Physical Geography. Translated by Zhang Tingquan and Hu Genkang*. Beijing: Foreign Languages Press, 1985.

Reuss, Martin. "Seeing Like an Engineer: Water Projects and the Mediation of the Incommensurable." *Technology and Culture* 49, 3 (2008): 531–46.

Richthofen, Ferdinand Paul Wilhelm Freiherr von. *Baron Richthofen's Letters, 1870–1872*. Shanghai: North China Herald, n.d.

Rowe, William. *Hankow: Commerce and Society in a Chinese City, 1796–1889*. Stanford: Stanford University Press, 1984.

———. *Hankow: Conflict and Community in a Chinese City, 1796–1895*. Stanford: Stanford University Press, 1989.

Roxby, Percy M. "Wu-Han: The Heart of China." *Scottish Geographical Magazine* 32 (1916): 266–79.

Sawyer, W. A. "Hookworm Disease as Related to Industry in Australia." *The American Journal of Tropical Medicine* 327, 3 (1923): 159–76.

Schoppa, R. Keith. *Chinese Elites and Political Change: Zhejiang Province in the Early Twentieth Century*. Cambridge: Harvard University Press, 1982.

Scott, James. *The Moral Economy of the Peasant: Rebellion and Subsistence in Southeast Asia*. New Haven: Yale University Press, 1976.

———. *Weapons of the Weak: Everyday Forms of Peasant Resistance*. New Haven: Yale University Press, 1985.

———. *Domination and the Arts of Resistance: Hidden Transcripts*. New Haven: Yale University Press, 1990.

———. *Seeing Like a State: How Certain Schemes to Improve the Human Condition Have Failed*. New Haven: Yale University Press, 1998.

Shaffer, Lynda. "Anyuan: The Cradle of the Chinese Workers' Revolutionary Movement, 1921–1922." In *Columbia Essays in International Affairs*, Vol.

5, 166–201. Ed. Andrew W. Cordier. New York: Columbia University Press, 1969.

———. *Mao and the Workers: The Hunan Labor Movement, 1920–1923*. Armonk: M. E. Sharpe, 1982.

Sheng Xuanhuai. *Yujai congao* (The collected works of Sheng Xuanhuai). Taibei: Wenhai chubanshe, 1974.

Skinner, G. William. "Marketing and Social Structure in Rural China," Parts I and II. *Journal of Asian Studies* 24, 1 and 2 (1965): 3–43, 195–228.

———. "Presidential Address: The Structure of Chinese History." *Journal of Asian Studies* 44, 2 (1985): 271–92.

———. "Sichuan's Population in the Nineteenth Century: Lessons from Disaggregated Data." *Late Imperial China* 8, 1 (1987): 1–79.

Smith, Carol, ed. *Regional Analysis: Vol. I: Economic Systems*. New York: Academic Press, 1976.

———, ed. *Regional Analysis: Vol. II: Social Systems*. New York: Academic Press, 1976.

Smith, Wilfred. *A Geographical Study of Coal and Iron in China*. Liverpool: University Press of Liverpool, 1926.

Statham, I. C. F. *Coal Mining Practice Four Volumes*. London: Caxton Publishing Company, 1958.

Stearns, Peter N. *The Industrial Revolution in World History*. Boulder: Westview Press, 1998.

Steinhardt, Nancy Shatzman. *Chinese Imperial City Planning*. Honolulu: University of Hawaii Press, 1990.

Sun E-tu Zen. "Mining Labor in the Ch'ing Period." In *Approaches to Modern Chinese History*, 45–67. Ed. Albert Feuerwerker. Berkeley: University of California Press, 1967.

———."Ch'ing Government and the Mineral Industries before 1800." *Journal of Asian Studies* 27, 4 (1968): 835–45.

Swensen, K. P. "The Pinghsiang Colliery." *Mining and Scientific Press* (October 29, 1910): 564–67.

Taussig, Michael T. *The Devil and Commodity Fetishism in South America*. Chapel Hill: University of North Carolina Press, 1980.

Tegengren, F. R. *The Iron Ores and Iron Industry of China, Part II*. Beijing: Geological Survey of China, Ministry of Agriculture and Commerce, 1923–1924.

Teng, Ssu-yu, and John K. Fairbank, eds. *China's Response to the West: A Documentary Survey 1839–1923*. New York: Antheneum, 1970.

Thompson, E. P. *The Making of the English Working Class*. New York; Vintage, 1963.

Thompson, Elspeth. *The Chinese Coal Industry: An Economic History*. London: Routledge Curzon, 2003.

Ting, Leonard G. "The Coal Industry of China," Pts. I and II. *Nankai Social and Economic Quarterly* 10, 1 and 10, 2 (1937): 32–74 and 193–277.

Ting, V. K. "China's Mineral Resources." *Far Eastern Review* 15, 2 (1919): 80–82.

Torgasheff, Boris P. *Mining Labor in China*. Shanghai: Bureau of Industrial and Commercial Information, Ministry of Industry, Commerce and Labor, and National Government of the Republic of China, 1930.

Trotter, Joe William. *Coal, Class, and Color: Blacks in Southern West Virginia, 1915–32*. Urbana: University of Illinois Press, 1990.

Tsai, Shih-shan Henry. *The Eunuchs in the Ming Dynasty*. Albany: State University of New York Press, 1996.

Tyng, Walworth. "The Miners' Church at Peaceful Spring: Among Collieries and Coke Ovens at Anyuen: A Vivid Picture of Our Work in a Little Known Part of the District of Hankow." *The Spirit of the Missions* 90 (1925): 474–77.

Union Research Institute. *Who's Who in Communist China*. Hong Kong: Union Research Institute, 1969.

Van Onselen, Charles. *Chibaro: African Mine Labour in Southern Rhodesia, 1900–1933*. Johannesburg: Ravan Press, 1980.

———. *New Babylon New Nineveh: Everyday life on the Witwatersrand, 1886–1914*. Johannesburg: Jonathan Ball, 1982.

Wakeman, Frederick, Jr. *The Great Enterprise: The Manchu Reconstruction of Imperial Order in Seventeenth-Century China*. Berkeley: University of California Press, 1985.

Walker, Kathy Le Mons. *Chinese Modernity and the Peasant Path: Semicolonialism in the Northern Yangzi Delta*. Stanford: Stanford University Press, 1999.

Wales, Nym (Helen Foster Snow). *The Chinese Communists: Sketches and Autobiographies of the Old Guard*. Westport: Greenwood, 1972.

Wallace, Anthony F. C. *St. Clair: A Nineteenth-Century Coal Town's Experience with a Disaster-Prone Industry*. New York: Knopf, 1987.

Wang Xuezhuang and Zhou Qiuguang. "Ping Liu Li hui tang qiyi xi wen ban wei: Jian lun yi ming 'Wei Zongchuan chuan"' (Problems concerning historical materials on the Pingxiang-Liuyang-Liling Uprising: Ascertaining the falsification of two documents "The Biography of Wei Zongchuan"). *Lishi yenjiu* 5 (1989): 137–52.

Wen jiapu (Wen Family lineage documents). N.p., n.d.

Wen Tingshi. *Wen Yunge xiansheng quanji* (Collected works of Wen Tingshi). Taipei: Wenhai chubanshe, 1975.

Will, Pierre-Étienne, and R. Bin Wong, eds. *Nourish the People: The State and Civilian Granary System in China, 1650–1850*. Ann Arbor: University of Michigan Press, 1991.

Williams, E. T. *Recent Chinese Legislation Relating to Commercial, Railway, and Mining Enterprises with Regulations for Registration of Trade Marks, and for the Registration of Companies*. Shanghai: Shanghai Mercury, 1905.

Wolf, Eric R. *Peasant Wars of the Twentieth Century*. New York: Harper Torchbooks, 1969.

———. "On Peasant Rebellions." In *Peasants and Peasant Societies: Selected Readings*, 264–74. Ed. Teodor Shanin. New York: Penguin Books, 1971.

Woodhead, H. G. W., ed. *The China Yearbook, 1926–27*. Chicago: University of Chicago Press, 1927.

Wou, Orodic Y. K. "Development, Underdevelopment and Degeneration: The Introduction of Rail Transport into Honan." *Asian Profile* 7, 3 (1984): 215–30.

Wright, Tim. "A Method of Evading Management: Contract Labor in Chinese Coal Mines before 1937." *Comparative Studies in Society and History* 23, 4 (1981): 656–78.

———. *Coal Mining in China's Economy and Society, 1895–1937*. Cambridge: Cambridge University Press, 1984.

———. " 'The Spiritual Heritage of Chinese Capitalism': Recent Trends in the Historiography of Chinese Enterprise Management." *The Australian Journal of Chinese Affairs* 19, 20 (1988): 185–214.

———, ed. *The Chinese Economy in the Early Twentieth Century: Recent Chinese Studies*. New York: St. Martin Press, 1992.

Wu Chang-chuan. "Cheng Kuan-ying: A Case Study of Merchant Participation in the Chinese Self-Strengthening Movement (1878–1884)." PhD dissertation. Columbia University, 1974.

Wu Shellen Xiao. "Underground Empires: German Imperialism and the Introduction of Geology in China, 1860–1919." PhD dissertation. Princeton University, 2010.

Xi Rong, ed. *Pingxiang xianzhi* (Pingxiang County gazetteer). Taibei: Chengwen chubanshe, 1975; originally published 1873.

Xiao Ping and Liu Zhiyong. *Shi hua Pingxiang* (Poetry and pictures of Pingxiang). Beijing: Zhongguo wenhua chubanshe, 2006.

Xu Dixin and Wu Chengming, eds. *Chinese Capitalism, 1522–1840*. Houndmills: Macmillan, 2000.

Xue, Yong. " 'Fertilizer Revolution?' A Critical Response to Pomeranz's Theory of 'Geographic Luck.' " *Modern China* 33, 2 (2007): 195–229.

Yan Bing. "Heian de niandai: bu qu de douzheng" (Dark ages: Unyielding struggle). In *Lishi zhishi congshu: Anyuan lukuang gongren julebu shihua*. Nanchang: Jiangxi renmin chubanshe, 1983.

Yang Guozhen. *Ming Qing tudi qi yue wenshu yanjiu* (Study of collected land contracts of the Ming and Qing). Beijing: Renmin chubanshe, 1988.

Yoon, Seungjoo. "Constitutional Change in the Lower Echelon of the Qing Bureaucracy: The Formation, Reformation, and Transformation of Zhang Zhidong's Document Commissioners, 1885–1909." PhD dissertation. Harvard University, 1999.

Zelin, Madeline. *The Magistrate's Tael: Rationalizing Fiscal Reform in Eighteenth-Century Ch'ing China*. Berkeley: University of California Press, 1984.

———. "The Rise and Fall of the Fu-Rong Salt-Yard Elite: Merchant Dominance in Late Qing China." In *Chinese Local Elite and Patterns of Dominance*, 82–109. Ed. Joseph W. Esherick and Mary Backus Rankin. Berkeley: University of California Press, 1990.

———. *The Merchants of Zigong: Industrial Entrepreneurship in Early Modern China*. New York: Columbia University Press, 2005.

Zelin, Madeline, Jonathan K. Ocko, and Robert Gardella, eds. *Contract and Property in Early Modern China*. Stanford: Stanford University Press, 2004.

Zhang Yi, Zhu Xiangqing, and Fan Yinfei, eds. *Jiangxi sheng zhi* (Jiangxi provincial gazetteer). Beijing: Zhonggong zhong yang dang xiao chubanshe, 1993.

Zunz, Olivier. *Making America Corporate, 1870–1920*. Chicago: University of Chicago Press, 1990.

Index